AF760439

Advances in Industrial Control

Series editors

Michael J. Grimble, Glasgow, UK
Michael A. Johnson, Kidlington, UK

More information about this series at http://www.springer.com/series/1412

Remigiusz Wiśniewski

Prototyping of Concurrent Control Systems Implemented in FPGA Devices

Springer

Remigiusz Wiśniewski
Faculty of Computer, Electrical and Control Engineering
Institute of Electrical Engineering, University of Zielona Góra
Zielona Góra
Poland

ISSN 1430-9491 ISSN 2193-1577 (electronic)
Advances in Industrial Control
ISBN 978-3-319-45810-6 ISBN 978-3-319-45811-3 (eBook)
DOI 10.1007/978-3-319-45811-3

Library of Congress Control Number: 2016949610

© Springer International Publishing Switzerland 2017
This work is subject to copyright. All rights are reserved by the Publisher, whether the whole or part of the material is concerned, specifically the rights of translation, reprinting, reuse of illustrations, recitation, broadcasting, reproduction on microfilms or in any other physical way, and transmission or information storage and retrieval, electronic adaptation, computer software, or by similar or dissimilar methodology now known or hereafter developed.
The use of general descriptive names, registered names, trademarks, service marks, etc. in this publication does not imply, even in the absence of a specific statement, that such names are exempt from the relevant protective laws and regulations and therefore free for general use.
The publisher, the authors and the editors are safe to assume that the advice and information in this book are believed to be true and accurate at the date of publication. Neither the publisher nor the authors or the editors give a warranty, express or implied, with respect to the material contained herein or for any errors or omissions that may have been made.

Printed on acid-free paper

This Springer imprint is published by Springer Nature
The registered company is Springer International Publishing AG
The registered company address is: Gewerbestrasse 11, 6330 Cham, Switzerland

Series Editors' Foreword

The series *Advances in Industrial Control* aims to report and encourage technology transfer in control engineering. The rapid development of control technology has an impact on all areas of the control discipline. New theory, new controllers, actuators, sensors, new industrial processes, computer methods, new applications, new design philosophies…, new challenges. Much of this development work resides in industrial reports, feasibility study papers and the reports of advanced collaborative projects. The series offers an opportunity for researchers to present an extended exposition of such new work in all aspects of industrial control for wider and rapid dissemination.

So very often, "industrial control" is taken to be synonymous with "process control", a domain perceived to be dominated by continuous-time processes. This identification is possibly reinforced by so many undergraduate control courses starting from the description and analysis of such systems with first- and second-order dynamical models. In reality, almost all continuous-time systems have to be switched on and off so these systems are embedded in a framework of conditional logic and are really hybrid systems. Looking just a little wider it is not difficult to find examples, indeed whole fields of systems, that are purely discrete. Nowhere is this more true than in the field of consumer electronics and digital control systems. Key components of these electronic systems are microprocessors, and a wide variety of programmable logic devices the latter evolving out of transistor-based application-specific integrated circuits. Given the importance of these devices it is a little surprising not to find this applications area having more profile in the activities and publications of the control community. As a corrective to this, the Series Editors of the *Advances in Industrial Control* monograph series are pleased to welcome this contribution *Prototyping of Concurrent Control Systems Implemented in FPGA Devices* by Remigiusz Wiśniewski, of the Institute of Electrical Engineering, University of Zielona Góra, Poland.

Dr. Wiśniewski's monograph opens with useful illustrations of the differences between sequential operations and concurrent operations before developing the major theme of the volume, concurrent control systems. The early chapters introduce two mathematical tools that are going to be instrumental in developing the prototyping methods. First is the theory and notation of Petri nets (Chap. 2) where the concepts of interpreted Petri nets are explained and illustrated. This is followed by graph theory (Chap. 3) and hypergraph theory (Chap. 4) with some original properties and theorems attributed to the author being presented. These are the mathematical tools that Dr. Wiśniewski uses in the remainder of the monograph to devise prototyping methods for concurrent control systems. The ultimate objective of this development phase is the application to programmable devices and in particular field programmable gate array devices. As the monograph progresses, Dr. Wiśniewski uses various easily understood real-world examples to illustrate concepts and techniques. These include a milling machine application, traffic lights, a beverage production process, and a smart home application. These examples help to leaven the use of the rather abstract mathematical tools that the modeling and analysis of discrete systems demands and makes the monograph an attractive addition to the *Advances in Industrial Control* series.

As has already been mentioned, monograph contributions to the discrete system's literature are not very frequent, but the reader may also find these *Advances in Industrial Control* monographs of interest:

- *Modelling and Analysis of Hybrid Supervisory Systems* by Emilia Villani, Paulo E. Miyagi and Robert Valette (ISBN 978-1-84628-650-6, 2007);
- *Deadlock Resolution in Automated Manufacturing Systems* by ZhiWu Li and MengChu Zhou (ISBN 978-1-84882-243-6, 2008); and, from the *Advanced Textbooks in Control and Signal Processing* series:
- *Modelling and Control of Discrete-event Dynamic Systems* by Branislav Hrúz and MengChu Zhou (ISBN 978-1-84628-872-2, 2007).

Industrial Control Centre
University of Strathclyde
Glasgow, Scotland, UK

Michael J. Grimble
Michael A. Johnson

Preface

This book is an attempt to overcome the gap between science and practice in the field of concurrent control systems specified by Petri nets. It combines theoretical aspects of concurrent systems (with the reference to algorithms and their computational complexity) supplemented with practical implementation and reconfiguration of a given system in an FPGA device.

We indented this book to be useful to CAD researchers, engineers, and designers of concurrent systems. The content of the book includes theoretical background and practical applications, especially regarding implementation and partial reconfiguration of FPGAs. The book may also be useful for students of electrical engineering, computer science, and discrete mathematics.

Almost all of the proposed algorithms and methods were implemented within the system *Hippo* developed at the University of Zielona Góra. Some of ready-to-use tools are available online at: www.hippo.iee.uz.zgora.pl.

I am grateful to:

- M. Wiśniewska for her love, exceptional support, and invaluable patience;
- A. Karatkevich for the support on almost all topics of the book, including verification of algorithms, theorems, and proofs;
- M. Adamski for fruitful discussions and for the inspiration regarding hypergraphs and perfect graphs;
- G. Benysek for the support on the preparation of this book;
- I. Grobelna for the perfect cooperation on the field of concurrent systems;
- G. Bazydło and G. Łabiak for the support and valuable discussions;
- L. Titarenko and A. Barkalov for the verification of the book content;
- M. Szajna for verifying English.

The results presented in Chap. 4 were obtained in cooperation with M. Wiśniewska and M. Adamski. Examples of the milling machine and smart home system are elaborated by I. Grobelna.

Zielona Góra
June 2016

Remigiusz Wiśniewski

Contents

Chapter 1
Introduction

1.1 Context and Motivation

Control systems surround us everywhere. They can be found in almost all areas of human life, such as medical care [4, 24, 26], transportation and automotive [19, 122, 129], artificial intelligence and robotics [36, 37, 83, 93], manufacturing [25, 64, 77], data and process mining [23, 53, 82, 134], digital devices and embedded systems [2, 34, 56, 80, 89, 106, 117, 142, 151], banking [6, 40, 115, 130], security and safety [49, 90, 91, 105].

Formally, a *control system* is a system, that is responsible for control and management of another system (often called as an *operational system*) [43, 52, 142]. Based on the set of input values (I) and the set of logic conditions (X), the control system sends proper control signals (Y) to the operational system. Additionally, a set of output values (O) is generated. The sets of inputs and outputs are used for communication with the environment of the whole system [142], however the operational part may also react on the external input signals (marked as *Data*) and produce output values (denoted as *Results*). The basic concept of the control and operational systems is illustrated in Fig. 1.1.

In the *sequential control systems*, the subsequent sets of operations for the operational system are computed and generated in a *sequence*. For example, a *central processor unit* (CPU) operates in this way, managing instructions to be handled [101, 126]. The most popular model of sequential control systems nowadays is a *finite-state machine* (FSM), also known as a finite-state automaton [8, 9, 39, 52, 148]. The typical FSM is a model of behavior that consists of a set of states, a set of transitions between the states, and a set of actions [9, 142].

Let us demonstrate the idea of sequential computation. An *RSA algorithm* is one of the most popular public-key cryptosystems [114]. In the initial stages of the method, the product of the two primes $n = p * q$ and Euler totient function $\phi = \phi(p) * \phi(q) = (p - 1) * (q - 1)$ are calculated [114]. Figure 1.2a presents a pseudo flowchart, where particular instructions are performed sequentially. Initially,

© Springer International Publishing Switzerland 2017
R. Wiśniewski, *Prototyping of Concurrent Control Systems Implemented in FPGA Devices*, Advances in Industrial Control,
DOI 10.1007/978-3-319-45811-3_1

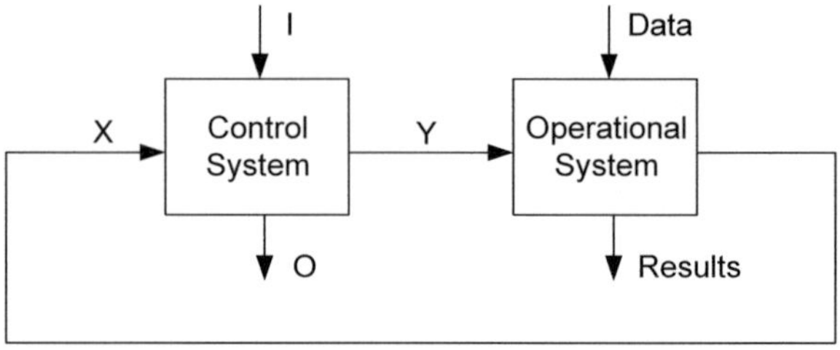

Fig. 1.1 Model of the control and operational systems

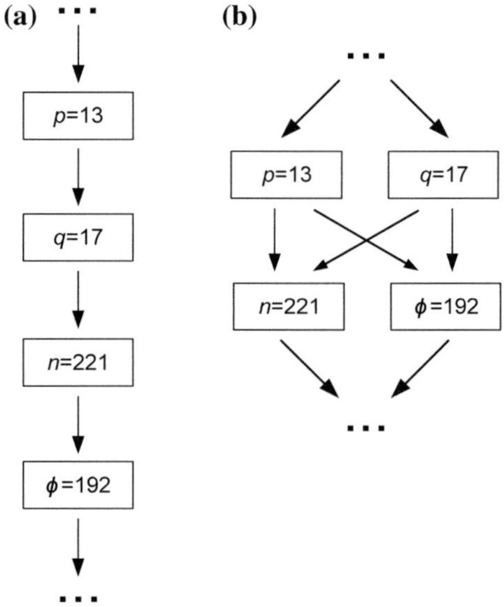

Fig. 1.2 An idea of **a** sequential and **b** concurrent computation

the first prime (p) is loaded (read) to the the control system. Analogically, the second prime (q) is loaded when the previous instruction is completed. Furthermore, the product of two primes is obtained. In the fourth step, the Euler totient function is calculated. Please note, that we just show the idea of sequential computation, thus all the presented actions are simplified and treated as a single instruction.

A *concurrent control system* can be seen as an extension of a sequential control system. The sets of operations for the operational system (or systems) may be computed and generated either *sequentially* or *concurrently*.

Figure 1.2b shows the idea of concurrent computation. In the presented example, both primes used in the RSA algorithm are read at the same time. Furthermore, the computation of their product, as well as calculation of the Euler totient function is performed simultaneously. Of course, proper synchronization between particular actions ought to be assured. For example, the computation of the product of the primes should be done only if both numbers have been loaded to the concurrent control system. Since synchronization is a very important aspect of prototyping of concurrent control systems, we will discuss it in more detail later.

Generally, a concurrent system is composed of the set of processes (components) that are executed at the same time. Each of them usually performs a particular task (or tasks). Such a situation is mostly known from the *concurrent programming* [13], where the required solution is solved by *distributed* computers (microprocessors). However, modelling of the whole system involves several problems regarding shared resources or synchronization of concurrent components. The classic design principle was initially formulated in [45, 59, 61], with further extensions oriented on particular language or prototyped system [3, 7, 13, 46, 67, 113, 124, 140].

In case of digital devices, a concurrent control system is often formally specified by a *Petri net*. This mathematical tool permits for easy and comfortable description of the prototyped design. First proposed in [109], Petri nets are still being developed and used in a various fields of science and industrial applications [41, 63, 79, 95, 97, 107, 123, 133, 137, 152].

Very often (but not always) prototyping of concurrent controllers specified by a Petri net involves splitting of the system into separate subsystems [35–37, 93, 97, 117]. If so, the control system is *decomposed* into a set of *modules*. Loosely speaking, decomposition usually divides the system into separate sequential components [57, 70, 97, 141]. The obtained modules can be implemented independently in various devices or digital systems [36, 37, 93, 117]. Moreover, decomposition is used in encoding of states of logic controllers implemented with programmable devices [22, 32, 33, 103, 128, 151].

The commonly used decomposition methods of concurrent control systems base on a linear algebra (place invariants analysis) [81, 84, 85, 132, 144, 149]. The fundamental computation technique can be found in [85], where as authors stated *a simple and fast* algorithm to obtain all invariants in a Petri net was proposed. Indeed, the presented technique is relatively fast and can be successfully applied to most concurrent systems described by Petri nets. However, there are serious limitations of methods that use linear algebra to decompose the concurrent system. First of all, the number of invariants may be exponential [31, 85]. It means that some systems described by Petri nets cannot be decomposed with the above technique because the algorithm is not able to find the solution in a reasonable time [144]. Furthermore, the obtained components ought to be verified in order to eliminate spurious solutions that may occur during the computation of invariants [118, 119, 144]. Finally, linear algebra technique searches for all the possible decomposition variants of the concurrent system, thus an additional selection of achieved modules has to be performed [120, 121, 141].

Beside the decomposition of the system, linear algebra is mainly used in the analysis of the concurrent controller described by a Petri net [76, 81, 88, 104, 118, 119]. First of all, this process permits to avoid redundancy (unreachable states), deadlocks, and reinitialization of the operation during its execution in the prototyped system [70, 118]. So-called *liveness* and *safeness* are crucial properties that are examined during the analysis of the concurrent controller specified by a Petri net [11, 17, 18, 71]. Furthermore, concurrency and sequentiality relations can be checked.

Those relations are especially useful in the selection of concurrent (or sequential) areas of the system.

Similarly to the decomposition, the main bottleneck of the analysis is exponential number of states that may occur in the design. Therefore, very often additional methods are applied prior to the main analysis of the system. For example, *reduction* techniques simplify the initial Petri net [14–16, 70, 97, 108, 150], while formation of the reduced set of reachable states permits to save time and space required for the analysis [14–16, 30, 65, 70, 73, 97, 108, 131, 135, 151]. However, very often formation of the reduced graph does not preserve the full information about the concurrency relation in the prototyped system [70]. Moreover, the size of the reduced graph can be still too large to be computed [50].

From the prototyping point of view, concurrent control systems can be divided into two main groups:

- *Distributed concurrent control systems*, generally implemented in more than one device. For example, it can be a set of connected *Programmable Logic Controllers* (PLCs), each of devices performs a sequential task [27, 72, 78, 138], a computer cluster (a set of connected computers distributed over the world [20, 110]), or a composition of various devices (for example, a composition of a microcontroller and a programmable device). Usually (but not always), each of the devices that composes the distributed system works in a different time domain, that is, each device uses its own clock signal [54, 127]. Therefore, distributed devices ought to be synchronized in order to work properly [55, 57, 74, 117].
- *Integrated concurrent control systems*, generally implemented in a single device, that supports concurrency (for example, digital circuits such as *Application Specific Integrated Circuits* (ASICs) [86, 142] or *Programmable Logic Devices* (PLDs) [1, 68, 69, 87, 142, 146]). Since the controller is implemented in a single device, its modules usually share the same clock signal, thus the synchronization between them is much easier than in case of distributed systems.

Various synchronization techniques of concurrent systems can be found in the literature. Among others, the general synchronization concepts are presented in [32, 34, 57, 58, 62, 74, 96, 127, 148]. An interesting idea addressed for distributed systems is shown in [96], where a controller is prototyped as a *globally asynchronous locally synchronous* (GALS) system. The proposed synchronization technique assures proper functionality of the decomposed modules that are working in different *time domains* (which means that decomposed modules are implemented in devices oscillated by various clock signals). Another method can be found in [57], where modifications in the structure of synchronized components or application of additional synchronization modules are proposed. A general synchronization technique oriented on a *model driven architecture* (MDA) is presented in [34]. Since MDA is a universal modelling technique [99], the concept of the *communication channels* [29] can be applied either to the distributed or integrated system [34].

Integrated concurrent control systems are often oriented on the implementation in the programmable device. Especially *Field Programmable Gate Arrays* (FPGAs) are considered. High performance, flexibility, and configurability resulted in their application in various aspects of human life, such as medicine [102, 125], cryptology [28, 44, 48, 98, 100], aerospace engineering [111], image processing [12, 38, 66], reconfigurable computing [60], measurement [94] and—of course—in concurrent control systems [2, 21, 42, 92, 139].

Latest FPGA devices, apart from the traditional static configuration of the entire system, offer much more sophisticated mechanisms. *Partial reconfiguration* allows replacement of selected parts of the system, without having to reprogram the entire structure of the FPGA [5, 147]. The system may be partially reconfigured in two ways: statically or dynamically. Especially, the second option seems to be very interesting, since it permits to replace a portion of the device without stopping it. It means, that during the *dynamic partial reconfiguration* a part of the controller can be modified, while the rest of the system is still running. However, preparation of the system intended for further partial reconfiguration requires a different approach compared to the traditional prototyping flow of integrated concurrent controllers. The reconfigurable area of the system ought to be selected carefully with paying special attention to the shared resources. Unfortunately, since partial reconfiguration is a relatively new idea, there is a lack of publications that formulate a precise prototyping flow intended for the concurrent control systems implemented in an FPGA with a possibility of further static or dynamic reconfiguration. The existing flows are rather dedicated for the particular exemplary controllers or involve additional specialized computer-aided tools to the prototyping process [10, 21, 47, 51, 75, 112, 116, 136, 143, 145].

Summarizing the above presented discussion, it can be stated that there are gaps in the existing methods of decomposition, analysis, and prototyping of concurrent control systems. Looking in more detail, two major aspects can be noticed:

- The main bottleneck of the decomposition and analysis of a concurrent control system specified by a Petri net is computational complexity of algorithms. In most cases such complexity is exponential, which means that the solution may never be found. Thus, the existing methods balance between optimal results and reasonable computational time.
- There is a lack of a dedicated prototyping flow of concurrent control systems specified by a Petri net intended for implementation in an FPGA and further partial reconfiguration of the design.

The aim of the book is to introduce concurrent control systems and to discuss solutions of effective prototyping, analysis, and decomposition of such systems. In the book, classical approaches based on linear algebra, as well as novel algorithms (with application of perfect graphs or hypergraphs) are analyzed and discussed in detail. Especially, the presented decomposition method based on the perfect graphs theory (Chaps. 3 and 6) may be very useful, since it offers obtaining exact results in polynomial time. On the other hand, not all systems can be decomposed with this algorithm, thus an alternate method based on the hypergraph theory is proposed as well. Furthermore, application of hypergraphs in the analysis of concurrency and sequentiality is proposed. Hypergraphs offer unique properties of the prototyped system in comparison to the traditional invariants analysis (Chaps. 4 and 5).

The main purpose of the book is to present prototyping methods of concurrent control systems, intended for implementation in FPGAs with a possibility of further partial reconfiguration of the controller. A concurrent system is represented by an *interpreted Petri net*, which naturally reflects concurrent and sequential relationships of a modelled controller. Furthermore, the system is decomposed into sequential components. Three alternative decomposition methods are presented in the book (Chap. 6). Each of decomposed modules is described with the use of *hardware description languages* (HDLs) such as Verilog or VHDL in terms of logic synthesis and further implementation in reconfigurable programmable systems. The modelling methods of concurrent systems in HDLs are shown for further implementation of the design in reconfigurable programmable devices (Chap. 8).

Finally, the complete prototyping flow of a concurrent control system intended for implementation in the FPGA with a possibility of further partial reconfiguration is proposed. Two different approaches are introduced. The first one is dedicated for the static partial reconfiguration of the design, while the second focuses on the dynamic partial reconfiguration of the concurrent control system (Chaps. 7 and 9).

1.2 Outline of the Book

The book is organized into ten chapters.

This chapter introduces concurrent control systems and elucidates the subject matter of the book.

Chapter 2 gives an overview of interpreted Petri nets as a specification of concurrent control systems. Computational complexity of algorithms is also briefly discussed.

Chapter 3 presents notations related to the graph theory. Furthermore, perfect graphs and comparability graphs are introduced. Author's theorems, proofs, and algorithms with detailed analysis of their computational complexity are presented.

Chapter 4 relates to the hypergraph theory. Beside well-known notations, new definitions, properties, theorems, and algorithms are introduced and analyzed in detail.

Chapter 5 deals with analysis of concurrent control systems. At the beginning, the most popular analysis method of the dynamic behavior of the net, based on the state equations and integer linear algebra (p-invariants computation) is presented. Furthermore, advanced methods of analysis of concurrency and sequentiality in a system are presented. New algorithms (especially related to the graph and hypergraph theories) are introduced.

Chapter 6 focuses on the decomposition of concurrent control systems. Three different decomposition methods are presented and compared in terms of their efficiency and computational complexity. Similarly to the previous chapter, algorithms are described in detail and illustrated by examples.

Chapter 7 gives prototyping techniques of concurrent systems. Algorithms and properties shown in previous chapters are applied in order to design a controller described by an interpreted Petri net. Initially, the classical approach, where a concurrent system is implemented as a single module, is presented. Further, a solution based on the decomposition of a controller is proposed.

Chapter 8 shows modelling methods of concurrent control systems. It is assumed that the system has been already prototyped with the use of ideas presented in Chap. 7. A controller is modelled as a finite-state machine and described with Hardware Description Languages (in particular Verilog HDL is used) for further logic synthesis and the system implementation in programmable devices.

Chapter 9 finalizes the prototyping flow of concurrent control systems. The description of programmable devices and Field Programmable Gate Arrays is presented. Furthermore, partial reconfiguration of a concurrent control system is introduced. Unique methods of static and dynamic partial reconfiguration of concurrent controllers implemented in an FPGA device are shown.

Chapter 10 concludes the prototyping methods shown in the book. The main contributions and results are summarized. Finally, other possible applications of the proposed methods and algorithms are suggested.

References

1. Altera homepage. http://www.altera.com. Accessed 4 Mar 2016
2. Adamski M, Karatkevich A, Wegrzyn M (eds) (2005) Design of embedded control systems. Springer, New York ISBN: 0-387-23630-9
3. Andrews GR (1991) Concurrent programming: principles and practice. Benjamin/Cummings Publishing Company
4. Andrews PS, Timmis J (2005) Inspiration for the next generation of artificial immune systems. In: Artificial immune systems. Springer, pp 126–138
5. Altera Stratix V FPGAs: ultimate flexibility through partial and dynamic reconfiguration. http://wl.altera.com/devices/fpga/stratix-fpgas/stratix-v/overview/partial-reconfiguration/stxv-part-reconfig.html. Accessed 11 Mar 2016
6. Attoui A (1997) An environment based on rewriting logic for parallel systems formal specification and prototyping. J Syst Archit 44(2):79–105
7. Banerjee A, Naumann DA (2002) Representation independence, confinement and access control [extended abstract]. In: ACM SIGPLAN notices, vol 37. ACM, pp 166–177
8. Baranov SI (1994) Logic synthesis for control automata. Kluwer Academic Publishers, Boston, MA
9. Barkalov A, Titarenko L (2009) Logic synthesis for FSM-based control units. Lecture notes in electrical engineering, vol 53. Springer, Berlin
10. Barkalov A, Wegrzyn M, Wiśniewski R (2006) Partial reconfiguration of compositional microprogram control units implemented on FPGAs. In: Proceedings of IFAC workshop on programmable devices and embedded systems (Brno), pp 116–119
11. Barkaoui K, Minoux M (1992) A polynomial-time graph algorithm to decide liveness of some basic classes of bounded Petri nets. In: Proceedings of the 13th international conference on application and theory of Petri nets, Sheffield, UK, pp 62–75
12. Batlle J, Martı J, Ridao P, Amat J (2002) A new FPGA/DSP-based parallel architecture for real-time image processing. Real-Time Imaging 8(5):345–356
13. Ben-Ari M (2006) Principles of concurrent and distributed programming. Pearson Education
14. Berthelot G (1986) Checking properties of nets using transformation. In: Advances in Petri Nets'85. Lecture notes in computer science, vol 222. Springer, pp 19–40

15. Berthelot G, Roucairol C (1976) Reduction of Petri nets. Mathematical foundations of computer science. Lecture notes in computer science, vol 45. Springer, Berlin, pp 202–209
16. Berthelot G, Roucairol C, Valk R (1980) Reduction of nets and parallel programs. In: Lecture notes in computer science, vol 84. Springer, pp 277–290
17. Best E (1987) Structural theory of Petri nets: the free choice hiatus. In: Lecture notes in computer science, vol 254. Springer, New York, pp 168–206
18. Best E, Thiagarajan P (1987) Some classes of live and safe Petri nets. In: Voss K, Genrich H, Rozenberg G (eds) Concurrency and Nets. Springer, Berlin, Heidelberg, pp 71–94
19. Bjørner D (2003) New results and trends in formal techniques and tools for the development of software for transportation systems-a review. In: Proceedings 4th symposium on formal methods for railway operation and control systems (FORMS03). L'Harmattan Hongrie, Budapest
20. Brewer EA (2000) Towards robust distributed systems. In: PODC, vol 7
21. Bukowiec A, Doligalski M (2013) Petri net dynamic partial reconfiguration in FPGA. In: Computer aided systems theory-EUROCAST. Springer, pp 436–443
22. Carmona J, Cortadella J (2006) State encoding of large asynchronous controllers. In: DAC, pp 939–944
23. Carmona J, Cortadella J, Kishinevsky M (2009) Divide-and-conquer strategies for process mining. Business Process Management. Lecture notes in computer science, vol 5701. Springer, Berlin, Heidelberg, pp 327–343
24. Chen M, Hofestädt R (2003) Quantitative Petri net model of gene regulated metabolic networks in the cell. In Silico Biol 3(3):347–365
25. Chen Y, Li Z, Al-Ahmari A (2013) Nonpure Petri net supervisors for optimal deadlock control of flexible manufacturing systems. IEEE Trans Syst Man Cybern Syst 43(2):252–265
26. Chinzei K, Hata N, Jolesz F, Kikinis R (2000) MR compatible surgical assist robot: system integration and preliminary feasibility study. In: Delp S, DiGoia A, Jaramaz B (eds) Medical image computing and computer-assisted intervention—MICCAI 2000. Lecture notes in computer science, vol 1935. Springer, Berlin, Heidelberg, pp 921–930
27. Chmiel M, Mocha J, Hrynkiewicz E, Polok D (2013) About implementation of IEC 61131–3 IL operators in standard microcontrollers. Program Devices Embed Syst 12:144–149
28. Chodowiec P, Gaj K (2003) Very compact FPGA implementation of the AES algorithm. In: Cryptographic hardware and embedded systems-CHES 2003. Springer, pp 319–333
29. Christensen S, Hansen ND (1994) Coloured Petri nets extended with channels for synchronous communication. In: Application and theory of Petri nets 1994, pp 159–178
30. Clarke EM, Grumberg O, Minea M, Peled DA (1999) State space reduction using partial order techniques. STTT 2(3):279–287
31. Colom JM, Silva M (1989) Convex geometry and semiflows in P/T nets. A comparative study of algorithms for computation of minimal p-semiflows. In: Advances in Petri nets 1990. Springer, pp 79–112
32. Cortadella J (2002) Logic synthesis for asynchronous controllers and interfaces., Springer series in advanced microelectronicsSpringer, Berlin, New York
33. Cortadella J, Kishinevsky M, Lavagno L, Yakovlev A (1998) Deriving Petri nets from finite transition systems. IEEE Trans Comput 47(8):859–882
34. Costa A, Barbosa P, Gomes L, Ramalho F, Figueiredo J, Junior A (2010) Properties preservation in distributed execution of Petri nets models. Emerg Trends Technol Innov 314:241–250
35. Costa A, Gomes L (2009) Petri net partitioning using net splitting operation. In: 7th IEEE international conference on industrial informatics (INDIN 2009). IEEE, pp 204–209
36. Costelha H, Lima P (2007) Modelling, analysis and execution of robotic tasks using Petri nets. In: International conference on intelligent robots and systems, pp 1449–1454
37. Costelha H, Lima P (2010) Petri net robotic task plan representation: modelling, analysis and execution. In: Kordic V (ed) Autonomous agents. InTech, pp 65–89
38. Crookes D, Benkrid K, Bouridane A, Alotaibi K, Benkrid A (2000) Design and implementation of a high level programming environment for FPGA-based image processing. In: IEE proceedings vision, image and signal processing, vol 147. IET, pp 377–384

39. Czerwinski R, Kania D (2012) Area and speed oriented synthesis of FSMs for PAL-based CPLDs. Microprocess Microsyst Embed Hardw Des 36(1):45–61
40. Dai L, Guo W (2007) Concurrent subsystem-component development model (CSCDM) for developing adaptive e-commerce systems. In: Computational science and its applications–ICCSA 2007. Springer, pp 81–91
41. David R, Alla H (2005) Discrete, continuous, and hybrid Petri nets. Springer
42. De Castro A, Zumel P, García O, Riesgo T, Uceda J (2003) Concurrent and simple digital controller of an AC/DC converter with power factor correction based on an FPGA. IEEE Trans Power Electron 18(1):334–343
43. De Micheli G (1994) Synthesis and optimization of digital circuits. McGraw-Hill, New York, NY
44. Deepakumara J, Heys HM, Venkatesan R (2001) FPGA implementation of MD5 hash algorithm. In: Canadian conference on electrical and computer engineering, 2001, vol 2. IEEE, pp 919–924
45. Dijkstra EW (1968) Cooperating sequential processes. Springer
46. Dijkstra EW (2001) Solution of a problem in concurrent programming control. In: Pioneers and their contributions to software engineering. Springer, pp 289–294
47. Doligalski M, Bukowiec A (2013) Partial reconfiguration in the field of logic controllers design. Int J Electron Telecommun 59(4):351–356
48. Eguro K, Venkatesan R (2012) FPGAs for trusted cloud computing. In: 2012 22nd international conference on field programmable logic and applications (FPL). IEEE, pp 63–70
49. Ellis A (2002) System and method for maintaining n number of simultaneous cryptographic sessions using a distributed computing environment, 9 Nov 2002. US Patent 6,484,257
50. Finkel A (1991) The minimal coverability graph for Petri nets. In: Advances in Petri nets 1993, papers from the 12th international conference on applications and theory of Petri nets, Gjern, Denmark, June 1991, pp 210–243
51. Fons F, Fons M, Cantó E, López M (2013) Real-time embedded systems powered by FPGA dynamic partial self-reconfiguration: a case study oriented to biometric recognition applications. J Real-Time Image Process 8(3):229–251
52. Gajski D (1996) Principles of digital design. Prentice Hall, Upper Saddle River, NJ
53. Goedertier S, Martens D, Vanthienen J, Baesens B (2009) Robust process discovery with artificial negative events. J Mach Learn Res 10:1305–1340
54. Gomes L, Costa A, Barros JP, Lima P (2007) From Petri net models to VHDL implementation of digital controllers. In: Industrial electronics society, 2007. IECON 2007. 33rd annual conference of the IEEE. IEEE, pp 94–99
55. Gomes L, Costa A, Barros JP, Moutinho F, Pereira F (2013) Merging and splitting Petri net models within distributed embedded controller design. In: Embedded computing systems: applications, optimization, and advanced design: applications, optimization, and advanced design, p 160
56. Gourcuff V, De Smet O, Faure J-M (2006) Efficient representation for formal verification of PLC programs. In: 8th international workshop on discrete event systems, pp 182–187
57. Grobelna I, Wiśniewski R, Grobelny M, Wiśniewska M (2016) Design and verification of real-life processes with application of Petri nets. IEEE Trans Syst Man Cybern Syst http://dx.doi.org/10.1109/TSMC.2016.2531673
58. Habermann AN (1972) Synchronization of communicating processes. Commun ACM 15(3):171–176
59. Hansen PB (1970) The nucleus of a multiprogramming system. Commun ACM 13(4):238–241
60. Hauck S, DeHon A (2010) Reconfigurable computing: the theory and practice of FPGA-based computation. Morgan Kaufmann
61. Hoare CAR (1972) Towards a theory of parallel programming. In: The origin of concurrent programming. Springer, pp 231–244
62. Hoare, CAR (1978) Communicating sequential processes. Springer

63. Holloway LE, Krogh BH (1990) Synthesis of feedback control logic for a class of controlled Petri nets. IEEE Trans Autom Control 35(5):514–523
64. Hu H, Zhou M, Li Z (2011) Supervisor design to enforce production ratio and absence of deadlock in automated manufacturing systems. IEEE Trans Syst Man Cybern Part A Syst Hum 41(2):201–212
65. Janicki R, Koutny M (1991) Using optimal simulations to reduce reachability graphs. In: Clarke EM, Kurshan RP (eds) Proceedings of the 2nd international conference on computer-aided verification CAV'90. LNCS, vol 531. Springer, London, pp 166–175
66. Johnston C, Gribbon K, Bailey D (2004) Implementing image processing algorithms on FPGAs. In: Proceedings of the eleventh electronics New Zealand conference, ENZCon'04, pp 118–123
67. Jones CB (2003) Wanted: a compositional approach to concurrency. In: Programming methodology. Springer, pp 5–15
68. Kania D (1999) Two-level logic synthesis on PAL-based CPLD and FPGA using decomposition. In: EUROMICRO conference, 1999. Proceedings. 25th, vol 1. IEEE, pp 278–281
69. Kania D, Kulisz J (2007) Logic synthesis for PAL-based CPLD-s based on two-stage decomposition. J Syst Softw 80(7):1129–1141
70. Karatkevich A (2007) Dynamic analysis of Petri net-based discrete systems. Lecture notes in control and information sciences, vol 356. Springer, Berlin
71. Kemper P (2004) O($|P||T|$)-algorithm to compute a cover of S-components in EFC-nets
72. Kim Y, Evans RG, Iversen WM (2008) Remote sensing and control of an irrigation system using a distributed wireless sensor network. IEEE Trans Instrum Meas 57(7):1379–1387
73. Klas G (1992) Hierarchical solution of generalized stochastic Petri nets by means of traffic processes. In: Jensen K (ed) Proceedingsof the 13th international conference on application and theory of Petri nets, Sheffield. Lecture notes in computer science, vol 616, pp 279–298
74. Krzywicki K, Andrzejewski G (2014) Data exchange methods in distributed embedded systems. In: New trends in digital systems design, pp 126–141. VDI Verlag GmbH, Düsseldorf
75. Łabiak G, Wegrzyn M, Muñoz AR (2015) Statechart-based design controllers for FPGA partial reconfiguration. In: XXXVI symposium on photonics applications in astronomy, communications, industry, and high-energy physics experiments (Wilga 2015), pp 96623Q–96623Q. International Society for Optics and Photonics
76. Lautenbach K (1986) Linear algebraic techniques for place/transition nets. In: Brauer W, Reisig W, Rozenberg G (eds) Advances in Petri nets. Lecture notes in computer science, vol 254. Springer, pp 142–167
77. Lewis R (1998) Programming industrial control systems using IEC 1131–3. IEE, London
78. Lewis RW (2001) Modelling distributed control systems using IEC 61499. Institution of Electrical Engineers, Stevenage, UK
79. Li H (1998) Petri net as a formalism to assist process improvement in the construction industry. Autom Constr 7(4):349–356
80. Li Z, Zhou M (2008) Control of elementary and dependent siphons in Petri nets and their application. IEEE Trans Syst Man Cybern Part A Syst Hum 38(1):133–148
81. Lin C, Tong Z (1991) An algorithm for computing S-invariants for high level Petri nets. Fachbericht Nr.15/91, Universität Koblenz-Landau, FB Informatik
82. Lynch NA (1983) Multilevel atomicity—a new correctness criterion for database concurrency control. ACM Trans Database Syst (TODS) 8(4):484–502
83. Ma S, Ding Y (2012) Application of neural network in the velocity loop of a gyro-stabilized platform. In: Lee G (ed) Advances in intelligent systems. Advances in intelligent and soft computing, vol 138. Springer, Berlin, Heidelberg, pp 355–361
84. Marinescu D, Beaven M, Stansifer R (1991) A parallel algorithm for computing invariants of Petri net models. In: Proceedings of PNets and performance models, IEEE Press
85. Martinez J, Silva M (1982) A simple and fast algorithm to obtain all invariants of a generalized Petri net. In: Selected papers from the European workshop on application and theory of Petri nets, London, UK. Springer, pp 301–310
86. Maxfield C (2004) The design warrior's guide to FPGAs. Academic Press Inc., Orlando, FL

87. Maxfield C (2009) FPGAs: world class designs: world class designs. Newnes
88. Memmi G, Roucairol G (1979) Linear algebra in net theory. In: Net theory and applications, proceedings of the advanced course on general net theory of processes and systems, Hamburg, 8–19 Oct 1979, pp 213–223
89. Milik A, Hrynkiewicz E (2012) Synthesis and implementation of reconfigurable PLC on FPGA platform. Int J Electron Telecommun 58(1):85–94
90. Mitchell JC (1998) Finite-state analysis of security protocols. In: Computer aided verification. Springer, pp 71–76
91. Miyamoto T, Nogawa H, Kumagai S et al (2006) Autonomous distributed secret sharing storage system. Syst Comput Jpn 37(6):55–63
92. Monmasson E, Cirstea MN (2007) FPGA design methodology for industrial control systems—a review. IEEE Trans Ind Electron 54(4):1824–1842
93. Montano L, García-Izquierdo F, Villarroel J (2000) Using the time Petri net formalism for specification, validation, and code generation in robot-control applications. Int J Rob Res 19:59–76
94. Moreno-Munoz A, Pallarés-López V, la Rosa D, González JJ, Real-Calvo R, González-Redondo M, Moreno-García I (2013) Embedding synchronized measurement technology for smart grid development. IEEE Trans Ind Inf 9(1):52–61
95. Moutinho F, Gomes L (2014) Asynchronous-channels within Petri net-based gals distributed embedded systems modeling. IEEE Trans Ind Inf 10(4):2024–2033
96. Moutinho F, Gomes L (2015) Distributed embedded controller development with Petri nets: application to globally-asynchronous locally-synchronous systems, 1st edn. Springer Publishing Company Incorporated
97. Murata T (1989) Petri nets: properties, analysis and applications. Proc IEEE 77:548–580
98. Nichols RK, Lekkas PC (2002) Wireless security. McGraw-Hill, New York
99. OMG Unified Modeling Language (OMG UML) Superstructure ver. 2.4.1. Object Management Group (2011)
100. Okada S, Torii N, Itoh K, Takenaka M (2000) Implementation of elliptic curve cryptographic coprocessor over GF (2m) on an FPGA. In: Cryptographic hardware and embedded systems—CHES 2000. Springer, pp 25–40
101. O'Regan G (2008) A brief history of computing. Springer Science & Business Media
102. Park J, Hwang J-T, Kim Y-C (2005) FPGA and ASIC implementation of ecc processor for security on medical embedded system. In: Third international conference on information technology and applications, 2005. ICITA 2005, vol 2. IEEE, pp 547–551
103. Pastor E, Cortadella J (1998) Efficient encoding schemes for symbolic analysis of Petri nets. In: DATE'98, pp 790–795
104. Pastor E, Roig O, Cortadella J, Badia RM (1994) Petri net analysis using boolean manipulation. In: Application and theory of Petri Nets'94, pp 416–435
105. Pathak A, Hu YC, Zhang M, Bahl P, Wang Y-M (2011) Fine-grained power modeling for smartphones using system call tracing. In: Proceedings of the sixth conference on computer systems. ACM, pp 153–168
106. Peng S, Zhou M (2004) Ladder diagram and Petri-net-based discrete-event control design methods. IEEE Trans Syst Man Cybern Part C Appl Rev 34(4):523–531
107. Peng SS, Zhou MC (2004) Ladder diagram and Petri-net-based discrete-event control design methods. IEEE Trans Syst Man Cybern Part C Appl Rev 34(4):523–531
108. Peterson JL (1981) Petri net theory and the modeling of systems. Prentice Hall PTR, Upper Saddle River, NJ, USA
109. Petri CA (1962) Kommunikation mit Automaten. Institut für Instrumentelle Mathematik, Schriften des IIM Nr. 2, Bonn
110. Piecuch P, Kucharski SA, Kowalski K, Musiał M (2002) Efficient computer implementation of the renormalized coupled-cluster methods: the R-CCSD [T], R-CCSD (T), CR-CCSD [T], and CR-CCSD (T) approaches. Comput Phys Commun 149(2):71–96
111. Pingree PJ (2010) Advancing NASA's on-board processing capabilities with reconfigurable FPGA technologies. INTECH Open Access Publisher

112. Quadri IR, Yu H, Gamatié A, Rutten E, Meftali S, Dekeyser J-L (2010) Targeting reconfigurable FPGA based socs using the UML MARTE profile: from high abstraction levels to code generation. Int J Embed Syst 4(3–4):204–224
113. Raynal M (2012) Concurrent programming: algorithms, principles, and foundations. Springer Science & Business Media
114. Rivest RL, Shamir A, Adleman L (1978) A method for obtaining digital signatures and public-key cryptosystems. Commun ACM 21(2):120–126
115. Santone A, Intilangelo V, Raucci D (2013) Application of equivalence checking in a loan origination process in banking industry. In: 2013 IEEE 22nd international workshop on enabling technologies: infrastructure for collaborative enterprises (WETICE). IEEE, pp 292–297
116. Shreejith S, Fahmy SA, Lukasiewycz M (2013) Reconfigurable computing in next-generation automotive networks. IEEE Embed Syst Lett 5(1):12–15
117. Silva E, Campos-Rebelo R, Hirashima T, Moutinho F, Malo P, Costa A, Gomes L (2014) Communication support for Petri nets based distributed controllers. In: 2014 IEEE 23rd international symposium on industrial electronics (ISIE), pp 1111–1116
118. Silva M (2013) Half a century after Carl Adam Petri's Ph.D. thesis: a perspective on the field. Ann Rev Control 37(2):191–219
119. Silva M, Terue E, Colom JM (1998) Linear algebraic and linear programming techniques for the analysis of place/transition net systems. Springer, Berlin, Heidelberg, pp 309–373
120. Stefanowicz L, Adamski M, Wiśniewski R (2013) Application of an exact transversal hypergraph in selection of SM-components. In: Technological innovation for the internet of things. Springer, Heidelberg, Dordrecht, pp 250–257
121. Stefanowicz L, Adamski M, Wiśniewski R, Lipiński J (2014) Application of hypergraphs to SMCs selection. In: Technological innovation for collective awareness systems. Springer, pp 249–256
122. Suk J, Lee Y, Kim S, Koo H, Kim J (2003) System identification and stability evaluation of an unmanned aerial vehicle from automated flight tests. KSME Int J 17(5):654–667
123. Szpyrka M, Matyasik P, Mrówka R, Kotulski L (2014) Formal description of Alvis language with α^0 system layer. Fundam Inf 129(1–2):161–176
124. Taboada GL, Ramos S, Expósito RR, Touriño J, Doallo R (2013) Java in the high performance computing arena: research, practice and experience. Sci Comput Program 78(5):425–444
125. Tanaka H, Ohnishi K, Nishi H, Kawai T, Morikawa Y, Ozawa S, Furukawa T (2009) Implementation of bilateral control system based on acceleration control using FPGA for multi-DOF haptic endoscopic surgery robot. IEEE Trans Ind Electron 56(3):618–627
126. Tanenbaum AS (2006) Structured computer organization. Pearson
127. Tanenbaum AS, Van Steen M (2007) Distributed systems. Prentice-Hall
128. Tkacz J, Adamski M (2012) Macrostate encoding of reconfigurable digital controllers from topological Petri net structure. Przeglad Elektrotechniczny 2012(8):137–140
129. Tzes A, Kim S, McShane W (1996) Applications of Petri networks to transportation network modeling. IEEE Trans Vehicular Technol 45(2):391–400
130. Unland R, Wanka U et al (1994) Mamba: automatic customization of computerized business processes. Inf Syst 19(8):661–682
131. Valmari A (1991) Stubborn sets for reduced state space generation. In: Advances in Petri nets 1990. Lecture notes in computer science, vol 483. Springer, Berlin, Germany, pp 491–515
132. Van Der Aalst W, Hee V (2004) Workflow management: models, methods, and systems. The MIT Press
133. Van der Aalst WM (1998) The application of Petri nets to workflow management. J Circ Syst Comput 8(01):21–66
134. van der Aalst WMP (2013) Decomposing Petri nets for process mining: a generic approach. Distrib Parallel Databases 31(4):471–507
135. Varpaaniemi K (1998) On the stubborn set method in reduced state space generation. Ph.D. thesis, Helsinki University of Technology, Department of Computer Science and Engineering

136. Vidal J, De Lamotte F, Gogniat G, Diguet J-P, Soulard P (2010) UML design for dynamically reconfigurable multiprocessor embedded systems. In: Proceedings of the conference on design, automation and test in Europe. European Design and Automation Association, pp 1195–1200
137. Voss K, Genrich HJ, Rozenberg G (2012) Concurrency and nets: advances in Petri nets. Springer Science & Business Media
138. Vyatkin V, I S of America (2007) IEC 61499 function blocks for embedded and distributed control systems design. ISA-Instrumentation, Systems, and Automation Society
139. Wegrzyn M, Adamski M, Karatkevich A, Muñoz AR (2014) FPGA-based embedded logic controllers. In: Proceedings of the 7th IEEE international conference on human system interactions, Lisbon, Portugal, pp 249–254
140. Williams A (2012) C++ concurrency in action. London
141. Wiśniewska M (2012) Application of hypergraphs in decomposition of discrete systems. Lecture notes in control and computer science, vol 23. University of Zielona Góra Press, Zielona Góra
142. Wiśniewski R (2009) Synthesis of compositional microprogram control units for programmable devices. Lecture notes in control and computer science, vol 14. University of Zielona Góra Press, Zielona Góra
143. Wiśniewski R, Barkalov A, Titarenko L (2008) Partial reconfiguration of compositional microprogram control units implemented on an FPGA. In: Proceedings of IEEE east-west design & test symposium-EWDTS, vol 8, pp 80–83
144. Wiśniewski R, Stefanowicz Ł, Bukowiec A, Lipiński J (2014) Theoretical aspects of Petri nets decomposition based on invariants and hypergraphs. Lecture notes in electrical engineering, Zhangjiajie, China, vol 308, pp 371–376
145. Wiśniewski R, Wiśniewska M, Adamski M (2016) Effective partial reconfiguration of logic controllers implemented in FPGA devices. In: Design of reconfigurable logic controllers. Springer, pp 45–55
146. Xilinx homepage. http://www.xilinx.com. Accessed 04 Mar 2016
147. Xilinx partial reconfiguration user guide. http://www.xilinx.com/support/documentation/sw_manuals/xilinx14_1/ug702.pdf. Accessed 11 Mar 2016
148. Yakovlev A, Gomes L, Lavagno L (2000) Hardware design and Petri nets. Springer
149. Zaitsev D (2004) Decomposition based calculation of Petri net invariants. In: Proceedings of the 25-th international conference on application and theory of Petri nets, pp 79–83
150. Zakrevskij A (1987) Verifying the correctness of parallel logical control algorithms. Program Comput Softw (USA) 13(5):218–221
151. Zakrevskij A, Pottosin Y, Cheremisinova L (2009) Design of logical control devices. TUT Press, Moskov
152. Zhou M (2012) Petri nets in flexible and agile automation, vol. 310. Springer Science & Business Media

Chapter 2
Related Work

Let us now formally specify a concurrent control system by an *Interpreted Petri net*, which is a powerful mathematical tool and naturally reflects concurrency in prototyped systems. Note, that the prototyping, analysis and decomposition methods presented in the book apply to the *integrated concurrent control systems* (cf. Chap. 1). Therefore, unless otherwise stated, we use the simple notations *concurrent control system*, *concurrent controller* in reference to this particular group.

2.1 Petri Nets and Interpreted Petri Nets

This section introduces basic definitions and notations regarding the Petri nets [5, 7, 8, 10, 13, 15, 17, 19, 23–25, 27–31, 33–35, 39].

Definition 2.1 A *Petri net* is a 4-tuple:

$$PN = (P, T, F, M_0), \tag{2.1}$$

where

- P is a finite set of places,
- T is a finite set of transitions,
- $F \subseteq (P \times T) \bigcup (T \times P)$ is a finite set of arcs,
- M_0 is an initial marking.

Set $P \cup T$ is a set of *nodes* of a Petri net.

© Springer International Publishing Switzerland 2017

R. Wiśniewski, *Prototyping of Concurrent Control Systems Implemented in FPGA Devices*, Advances in Industrial Control,
DOI 10.1007/978-3-319-45811-3_2

Definition 2.2 (*Marking*) State of a Petri net is called a *marking*, which can be seen as a distribution of tokens in the net places. If a place contains one or more tokens, it is called a *marked place*. A marking can be changed by means of *firing* (execution) of a transition.

Definition 2.3 (*Transition firing*) A transition t can be *fired* if every of its input places contains a token. Transition firing removes a token from every input place of t and adds a token to every output place of t.

An exemplary Petri net PN_1 (taken from [14, 37]) is shown in Fig. 2.1. The net contains six places $P = \{p_1, \ldots, p_6\}$ and three transitions $T = \{t_1, t_2, t_3\}$.

Definition 2.4 (*Input and output places*) Place p is an *input* place of transition t, if $(p, t) \in F$. Place p' is an *output* place of transition t, if $(t, p') \in F$. The set of input places of transition t is denoted by $\bullet t$, while $t\bullet$ denotes the set of output places of t.

Definition 2.5 (*Input and output transitions*) Transition t is an *input* transition of place p, if $(t, p) \in F$. Transition t' is an *output* transition of place p, if $(p, t') \in F$. The set of input transitions of place p is denoted by $\bullet p$, while $p\bullet$ denotes the set of output transitions of p.

Definition 2.6 (*Reachable marking*) A marking M' is *reachable* from marking M, if M' can be obtained from M by a finite sequence of transition firings. When marking M is not specified explicitly, a reachable marking of a net is understood as a marking reachable from its initial marking M_0.

A state of a Petri net is obtained by a distribution markers (tokens) on the places. Figure 2.2 illustrates all the possible markings of PN_1. There are three reachable markings in the presented net, that can be reached by consecutive firing of transitions t_1, t_2 and t_3. Changes between particular states for this net can be simply denoted as: $M_0 \xrightarrow{t_1} M_1 \xrightarrow{t_2} M_2 \xrightarrow{t_3} M_0$.

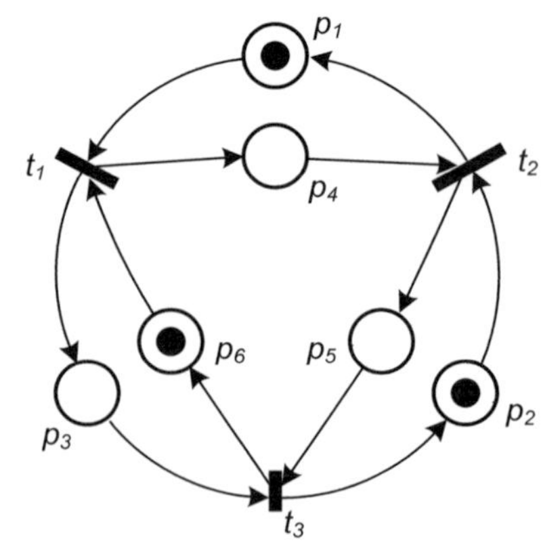

Fig. 2.1 An exemplary Petri net PN_1

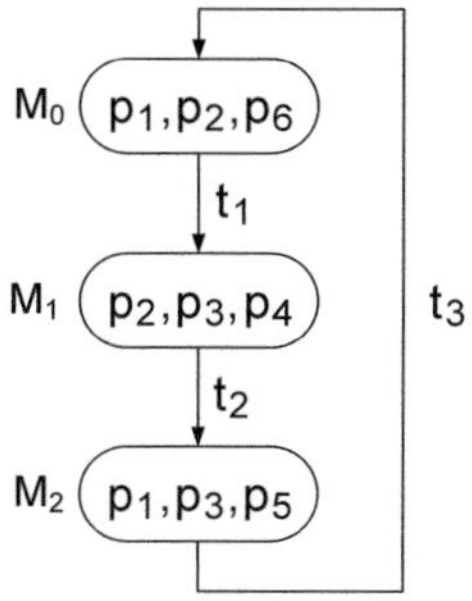

Fig. 2.2 All reachable markings of PN_1

The state M_0 consists of three places p_1, p_2 and p_6, which means that those places are initially marked. After firing of the transition t_1, tokens from its input places (p_1 and p_6) are moved to the output places (p_3 and p_4) of t_1 and the net reaches marking M_1. Please note that token at place p_2 remains unchanged. State M_1 enables transition t_2, which execution leads to the third marking M_2. At this state three places are marked: p_1, p_3 and p_5. Finally, firing of t_3 returns tokens into the initial marking M_0.

Definition 2.7 (*Enabled transitions in a given marking*) For a Petri net, transition t is *enabled in marking* M if $\forall p \in \bullet t : p \in M$, that is, all its input places are simultaneously marked in M. Any transition enabled in M can fire, changing the marking M to M'. $M[t>$ denotes that t is enabled in M, while $M[>$ indicates the set of all enabled transitions in M.

For PN_1, transition t_1 is enabled in M_0, which is denoted by $M_0[t_1>$. Furthermore, firing of t_1 changes an initial marking M_0 to M_1. Such a situation is represented as $M_0[t_1>M_1$. Similarly, $M_1[t_2>M_2$. Finally, $M_2[t_3>M_0$ moves tokens to the initial state.

Definition 2.8 (*Liveness*) A transition t is *live* if for every reachable marking M, t can be fired in M or in a marking reachable from M. A Petri net is *live* if all transitions in the net are live.

Definition 2.9 (*Safeness*) A place of a net is *safe* if there is no reachable marking such that the place contains more than one token. A Petri net is *safe* if each place in the net is safe.

The characterization of liveness and safeness has been studied by the researches all over the world. Fundamental theory regarding liveness and safeness of Petri nets was introduced in [6, 16]. Such analysis was extended in [15]. Finally, advanced algorithms allowing checking liveness and safeness of particular subclasses of Petri nets were proposed in [2–4, 21, 40].

Let us point out that analysis of liveness and safeness is out of the scope of this monograph. In our opinion, existing theorems and algorithms (especially those presented in [2–4, 15, 16, 21, 40]) exhausted this subject enough. Therefore, we decided to focus on the concurrency and sequentiality aspects of analysis of Petri nets.

Definition 2.10 (*Reversibility*) A Petri net is *reversible* if for each marking M, the initial marking M_0 is reachable from M. In other words, Petri net is reversible if it can always reach its the initial state.

Definition 2.11 (*Well-formed net*) A Petri net is *well-formed* if it is *live*, *safe* and *reversible*.

Definition 2.12 (*Conservativeness*) A Petri net is *conservative* if all the reachable markings contain the same number of tokens.

Definition 2.13 (*Pure net*) A Petri net is *pure* if it has no self-loops.

Definition 2.14 (*Path, strong connectedness*) A *path* in a Petri net PN is a sequence of nodes, connected by arcs. A net is *strongly connected* if for any pair (n_i, n_j) of its nodes there is a path leading from n_i to n_j.

Theorem 2.1 [3] *Let PN be a live and safe Petri net. Then PN is strongly connected.*

Petri net PN_1 is live, safe and conservative, therefore it is well-formed. Moreover, the net is pure and conservative, since all the markings contain exactly three tokens. PN_1 is strongly connected, because for any pair of nodes there is a path that connects them.

Definition 2.15 (*Interpreted Petri net*) An *interpreted Petri net* is a well-formed Petri net, defined as a 6-tuple:

$$PN = (P, T, F, M_0, X, Y), \tag{2.2}$$

where:

- P is a finite set of places,
- T is a finite set of transitions,
- $F \subseteq (P \times T) \bigcup (T \times P)$, is a finite set of arcs,
- M_0 is an initial marking,
- $X = \{x_1, x_2, \ldots, x_m\}$ is a binary vector of logic inputs,
- $Y = \{y_1, y_2, \ldots, y_n\}$ is a binary vector of logic outputs.

Interpreted Petri nets are very often used to specify the real-life controllers. The system communicates with the environment via input and output signals. The inputs are associated with transitions while its outputs are bounded to places. The transition is enabled if all the transition inputs are active (or condition tied to the transition is fulfilled). Therefore, input signals may preserve the net conflict-free. We shall show such a situation later in this section. Notice, that from the definition an interpreted

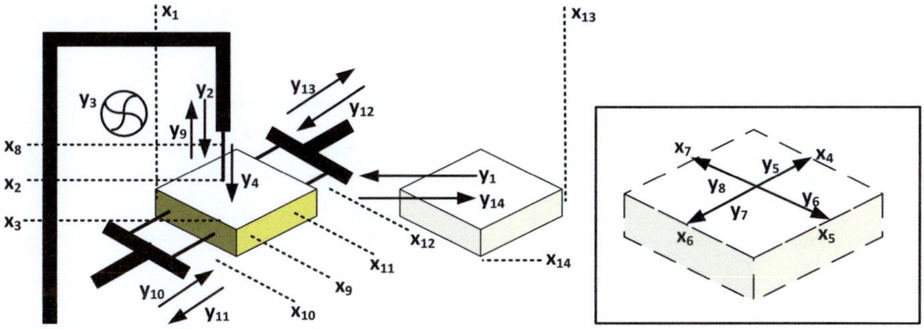

Fig. 2.3 A model of a milling machine

Petri net is well-formed, thus it is live, safe and reversible. It implies important additional properties. For example, based on the Theorem 2.1, each interpreted net is strongly connected.

Let us illustrate interpreted Petri nets by a real-life example. Figure 2.3 shows a modified system of the milling process, initially proposed in [38]. The main aim of the presented machine is to cut the square shapes from the wooden plank. The process is driven by a logic controller, specified by an interpreted Petri net PN_2, shown in Fig. 2.4.

There are 14 input signals denoted by $x_1, \ldots, x_{14}$, and 14 output signals marked as $y_1, \ldots, y_{14}$. Placement of a wooden plank on the tray (indicated by the sensor x_{14}) starts the whole process. First, the plank is moved (output signal y_1), until reached the right position (signalized by x_1). Simultaneously, a drill (y_2) is being set into the starting position (x_2). Next, the machine starts cutting the required shape from the wood (in the presented example—a square). The move of the drill is denoted by y_4 (immersion into the wood), y_5 (the drill moves to the right), y_6 (the drill moves to the down), y_7 (the drill moves to the left), y_8 (the drill moves to the top), y_9 (the drill goes up, to the initial position). Reaching the remaining positions is signalized by sensors x_3, x_4, x_5, x_6, x_7 and x_8, respectively. At the same time, while the shape is drilled, three concurrent actions are performed: a vacuum cleaner is turned on (y_3), and two assembly holes are drilled (signals y_{10}, y_{11} and y_{12}, y_{13}. Finally, the drilled shape is moved to the platform (y_{14}) to the tray. When the plank is taken away ($\overline{x_{14}}$), the system is ready for the further actions.

The net PN_2 consists of $|P| = 21$ places and $|T| = 17$ transitions. It is live, safe, reversible, and strongly connected. However, it is not conservative, since firing of t_1 moves and splits a single token from p_1 into p_2 and p_4.

Figure 2.5 shows another real-life example of a concurrent control system. The picture illustrates an advanced traffic lights system, driven by the FPGA device. There are three independent traffic lights that control particular lanes for cars (turning left, going straight, turning right). Additionally, two traffic lights for pedestrians are considered. In our consideration a simplified version of the controller (initially shown in [36]) will be presented. Assume, that all the car lanes operate in the same manner

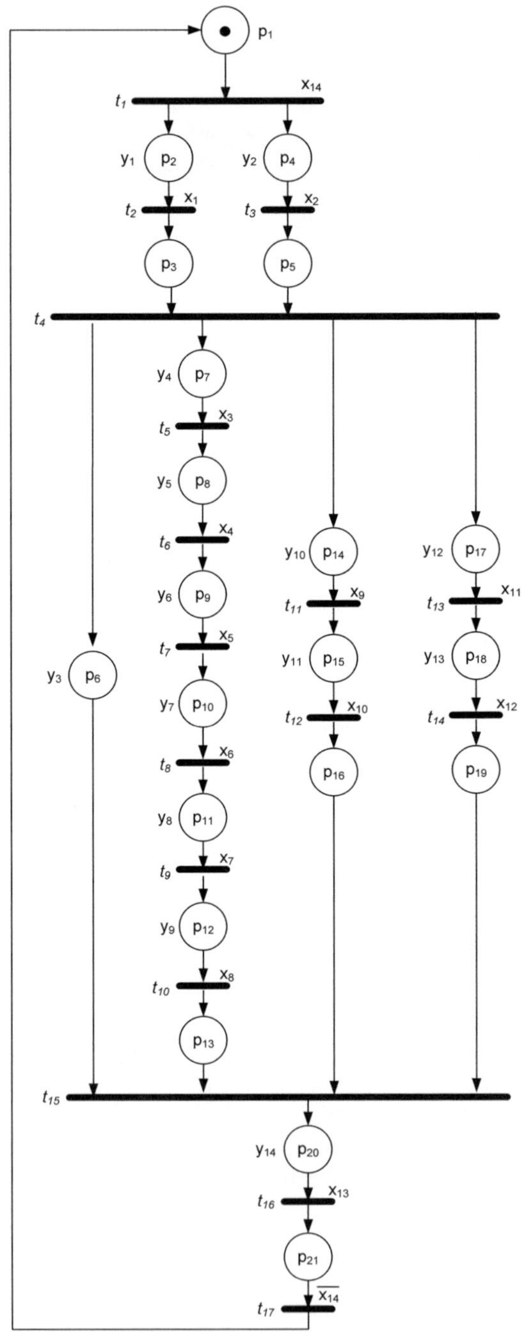

Fig. 2.4 An interpreted Petri net PN_2 that controls the milling machine

Fig. 2.5 Traffic lights system

and they are treated as a single traffic light. Similarly, both traffic lights for pedestrians are coupled. Such a system can be easily described by an interpreted Petri net (PN_3), as it is presented in Fig. 2.6. Initially, the red lights for cars (place p_4, output R_C) and for pedestrians (place p_6, output R_P) are active. If there is no request from pedestrians to cross the street (latched input signal req is inactive), the system enters the state, when the green light for cars (place p_2, output G_C) is shown (note, that the red light for pedestrians prevents collision). Such a situation takes place until a pedestrian wishing to cross the street pushes the button (signal req), which results in firing the transition t_2. Next, the yellow light for cars (place p_3, output Y_C) is flashed and the system goes back to the initial state. Since signal req is active, the transition t_4 is enabled and executed. The green light for pedestrians is shown to cross the street (it is assumed, that signal req is zeroed). The system returns to the initial state and the whole procedure is repeated.

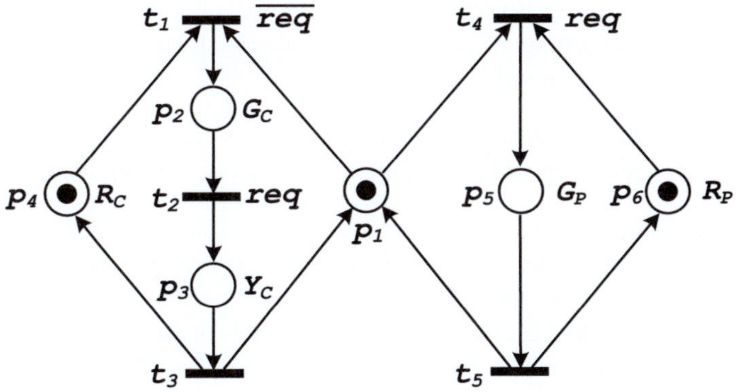

Fig. 2.6 An interpreted net PN_3 that controls the simplified version of traffic lights

Petri net PN_3 consists of $|P| = 6$ places and $|T| = 5$ transitions. It is live, safe, and reversible. The net is not conservative, because firing of t_1 consumes two tokens (from places p_1 and p_4), while only one token is moved to p_2.

Note that place p_1 has two output transitions: $p_1 \bullet = \{t_1, t_4\}$. Such a situation is called *conflict* in a Petri net. In case of interpreted Petri nets, it can be easily resolved by the input signals. In the particular example, signal req indicates, which transition is enabled. If there are no conflicts in the net, it is called *conflict-free* Petri net. For example, PN_1 is conflict-free, because each place has exactly one output transition.

It is assumed that interpreted Petri nets shown in this book are free of conflicts. It means that either the net is conflict-free, or such a conflict is resolved by the input signals. Furthermore, any Petri net that is live, safe and conflict-free can be classified as an interpreted Petri net. Thus, $PN_1 = (P, T, F, M_0, \emptyset, \emptyset)$ is an interpreted Petri net, but the sets of its input and output signals are empty: $X = Y = \emptyset$. Of course one may say that a controller specified by such a net does not have any sense, since there is no communication with the environment. Obviously it is true. However, such nets can be successfully applied for theoretical purposes or at those prototyping stages (mainly analysis, cf. Chap. 5), where input and output signals are not considered.

Definition 2.16 (*Concurrency in interpreted Petri nets*) Two places are *concurrent* if they are marked simultaneously at some reachable marking.

Definition 2.17 (*Sequentiality in interpreted Petri nets*) Two places are *sequential* if they cannot be marked simultaneously, i.e., there is no marking that contains both places.

It is assumed, that (in case of interpreted Petri nets) concurrency and sequentiality are complementary, i.e., particular place is either concurrent or sequential to the other place.

Definition 2.18 (*Reachability set, concurrency set*) *Reachability set* or *concurrency set* of a concurrent control system is the set of its reachable states (markings).

Reachability (concurrency) set, beside obvious concurrency analysis, can be also used to verify the correctness of the system behavior [11]. Popular representation of such a set is *reachability graph*.

Definition 2.19 (*Reachability graph*) *Reachability graph* of a concurrent control system is the directed graph of its reachable states, and the direct transitions between them.

Figure 2.2 (shown at the beginning of current chapter) presents an exemplary reachability graph for Petri net PN_1 (from Fig. 2.1).

Petri nets are classified according to their structure. In our consideration we use the following classification of a Petri net (taken directly from [25]):

1. State Machine (SM),
2. Marked Graph (MG),

3. Free-Choice net (FC-net),
4. Extended Free-Choice net (EFC-net),
5. Simple Net (SN), also known as Asymmetric Choice net (AC-net).

Let us define each of the presented subclasses.

Definition 2.20 (*State machine, SM-net*) A *state machine* is a Petri net for which every transition has exactly one input place and exactly one output place, i.e., $\forall t \in T : | \bullet t| = |t \bullet | = 1$.

Definition 2.21 (*Marked graph, MG-net*) A *marked graph* is a Petri net for which every place has exactly one input transition and exactly one output transition, i.e., $\forall p \in P : | \bullet p| = |p \bullet | = 1$.

Definition 2.22 (*Free-choice, FC-net*) A *free-choice net* is a Petri net for which every outgoing arc from a place is unique or is a unique incoming arc to a transition, i.e., $\forall p \in P : |p \bullet | \leq 1$ or $\bullet(p\bullet) = \{p\}$.

Definition 2.23 (*Extended free-choice net, EFC-net*) An *Extended Free-Choice net* is a Petri net for which every two places having a common output transition, have all their output transitions in common, i.e., $\forall p_i, p_j \in P : p_i \bullet \cap p_j \bullet \neq 0 \Rightarrow p_i \bullet = p_j \bullet$.

Definition 2.24 (*Simple net, SN*) A *Simple net* is a Petri net for which every two places having a common output transition, one of them has all the output transitions of the other (and possibly more), i.e.,$\forall p_i, p_j \in P : p_i \bullet \cap p_j \bullet \neq 0 \Rightarrow (p_i \bullet \subseteq p_j \bullet)$ or $(p_i \bullet \supseteq p_j \bullet)$.

All the nets, that do not belong to any of above subclasses are just classified as *Petri nets* (*PNs*). For the interpreted Petri nets, we shall use the same abbreviations for their subclasses. For example an interpreted Petri net that is classified as an EFC-net, will be shortly classified as an *interpreted EFC-net*.

Particular subclasses are structured as follow: SM and MG are simplest possible structures of a Petri net. FC is a generalization of SMs and MGs. It means that each net that belongs to SM is also classified as an FC-net. Similarly, each MG is an FC-net, as well. Furthermore, EFC is a generalization of FC, while SN is a generalization of EFC. Finally, PN is a generalization of SN. Figure 2.7 illustrates the hierarchy of subclasses of Petri nets.

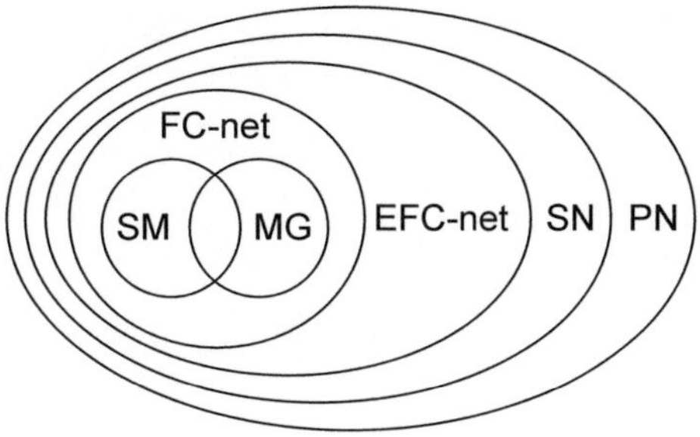

Fig. 2.7 Subclasses of Petri nets

Recall net PN_1, shown in Fig. 2.1. Every place of PN_1 has exactly one input transition and exactly one output transition (there are no multiple arcs outgoing from places). It means that such a net belongs to MG. Moreover, it is automatically classified as FC-net, EFC-net, SN, and PN, as well. Similarly, net PN_2 is also classified as MG-net, but PN_3 belongs to SN.

Classification of a Petri net may be useful during analysis of the net [25]. For example, liveness, safeness of some subclasses (SM, MG, FC, EFC) can be checked polynomially [2–4, 15, 16, 21].

Definition 2.25 (*State machine component*) A *state machine component (SMC, SM-component, S-component)* of a Petri net $PN=(P, T, F, M_0)$ is a Petri net $S=(P', T', F', M_0)$ such that:

1. S is an SM-net,
2. S is strongly connected,
3. $P' \subseteq P$,
4. $T' = \bullet P' \cup P'\bullet$,
5. $F' = F \cap (P' \times T') \cup (T' \times P')$,
6. S contains exactly one token in M_0.

Figure 2.8 shows all the state machine components that can be obtained in the net PN_1. There are four SMCs, consisting of the following places: $S_1 = \{p_1, p_4\}$, $S_2 = \{p_3, p_6\}$, $S_3 = \{p_2, p_5\}$, $S_4 = \{p_4, p_5, p_6\}$.

Definition 2.26 (*SM-decomposition*) *State machine decomposition (SM-decomposition, S-decomposition)* of a Petri net $PN = (P, T, F, M_0)$ is a set $\mathcal{S}$ of *elements* (often called *components* or *modules*) $\mathcal{S} = \{S_1, \ldots, S_n\}$ such that each place $p_i \in P$ is a place of exactly one component $S_j \in \mathcal{S}$. Each component S_j is an SM-net. If the particular place exists in more than one component, it is replaced by a place not belonging to P, called a *non-operational place* (*NOP*).

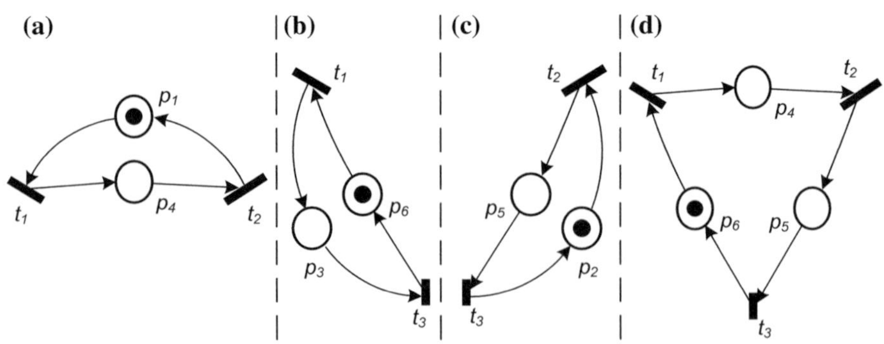

Fig. 2.8 All the SMCs of PN_1

Definition 2.27 (*Non-operational place, NOP*) Let PN be a Petri net and let $\mathcal{S}$ be a set of SMCs achieved during SM-decomposition of PN. If place $p \in P$ exists in more than one $S \in \mathcal{S}$, it is replaced by a *non-operational place* (*NOP*) in all $S \in \mathcal{S}$, but one. The set $P' = \{p_i, \ldots, p_j\}$ of places of the same component $S \in \mathcal{S}$ can be replaced by a single NOP if there exists a path in P' leading from p_i to p_j. Then, all transitions and arcs between p_i and p_j are removed, as well. A NOP is initially marked if any place substituted by it is initially marked.

Let us explain decomposition and NOPs by examples. Petri net PN_1 can be decomposed into three components: $\mathcal{S}$=$\{S_1, S_2, S_3\}$, where $S_1 = \{p_1, p_4\}$, $S_2 = \{p_3, p_6\}$, and $S_3 = \{p_2, p_5\}$. This is the only one possibility to decompose PN_1. There are no places that exist in more than one component, thus there is no need to apply non-operational places.

Consider Petri net PN_4 shown in Fig. 2.9a. There are six places in the net $P = \{p_1, \ldots, p_6\}$ and four transitions $T = \{t_1, \ldots, t_4\}$. The net is safe, live, and reversible. There are three SMCs in the PN_4: $S_1 = \{p_1, p_3, p_4\}$, $S_2 = \{p_2, p_3, p_4, p_6\}$, $S_3 = \{p_5, p_6\}$.

Figure 2.9b illustrates one of the possible decompositions of PN_4. In the presented example, two NOPs are applied, therefore the final set of decomposed components consists of the following places: $S_1 = \{p_1, p_3, p_4\}$, $S_2 = \{p_2, NOP_1, p_6\}$, $S_3 = \{p_5, NOP_2\}$.

The first non-operational place (NOP_1) is used in module S_2. It substitutes places p_3 and p_4 that already exist in the first component S_1. All the transitions and arcs

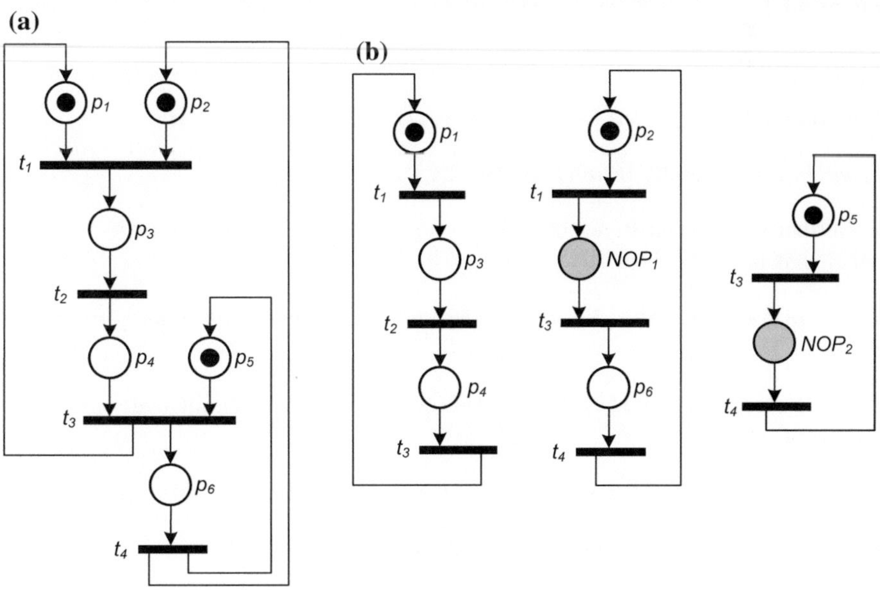

Fig. 2.9 Petri net PN_4 (**a**) and decomposed PN_4 (**b**)

belonging to the path connecting such places are removed as well. Furthermore, NOP_2 replaces p_6 in the third component S_3, since p_6 already exists in S_2.

Let us point out that there are more ways to decompose PN_4, depending on the application of non-operational places. For example, the alternative decomposition of the net may consists of the following components: $S_1 = \{p_1, NOP_1\}$, $S_2 = \{p_2, p_3, p_4, NOP_2\}$, $S_3 = \{p_5, p_6\}$, while NOP_1 substitutes places p_3, p_4, and NOP_2 replaces p_6.

More information about decomposition of Petri nets as well as decomposition algorithms can be found in Chap. 6.

Definition 2.28 (*SM-cover*) *State machine cover* (*SM-cover, S-cover*) of a Petri net $PN = (P, T, F, M_0)$ is a set $\mathcal{C}$ of state machine components $\mathcal{C} = \{S_1, \ldots, S_n\}$ such that each place $p_i \in P$ is a place of at least one component $S_j \in \mathcal{C}$.

SM-cover is very often confused with SM-decomposition. The main difference is that the particular place may exist in more than one component that belongs to the SM-cover. In opposite, the set of decomposed modules contains each place of the net exactly once. There are known conversion methods between SM-decomposition and SM-cover. Such a transformation can be done relatively easy, unless the conditions of the existence of SM-decomposition (or SM-cover) are not satisfied. More details can be found in [20].

Let us now introduce theorems regarding SM-decomposition and SM-cover. A very important relation between SM-cover and well-formed EFC-nets was shown in [9] (initially proposed for FC-nets in [15]):

Theorem 2.2 [9] *Well-formed EFC-nets are covered by SM-components.*

Since every interpreted EFC-net is well-formed, we immediately have

Theorem 2.3 *Interpreted EFC-nets are covered by SM-components.*

Proof Follows directly from Definition 2.15 and Theorem 2.2. □

Furthermore, the existence of SM-decomposition of a net depends on its safeness, which was proved in [20]:

Theorem 2.4 [20] *For a Petri net exists an SM-decomposition, if and only if PN is safe.*

Since every interpreted Petri net is safe by definition, we obtain a very important statement:

Theorem 2.5 *For an interpreted Petri net there always exists an SM-decomposition.*

Proof Follows directly from Definition 2.15 and Theorem 2.4. □

2.2 Computational Complexity of Algorithms

This section briefly introduces some preliminaries regarding computational complexity of algorithms. The presented notation we shall use during analysis of the computational complexity of algorithms shown in Chaps. 3–6.

The computational complexity presented in this book refers to the *time complexity*. Therefore, unless otherwise stated, we shall estimate the amount of *time* that is required in order to solve the problem. Note that only basic assumptions are presented. The detailed descriptions regarding computational complexity of algorithms can be found in [1, 12, 18, 22, 26, 32].

- *Time computational complexity of an algorithm* refers to the maximum number of interactions (steps) that the algorithm uses on a given size of a given input. Formally, *time complexity of algorithm* $\mathcal{A}$ is the function $f : \mathcal{N} \rightarrow \mathcal{N}$, where $f(n)$ is the maximum number of interactions (steps) that $\mathcal{A}$ uses on any input of length n [32].
- *Big-O notation* is used to estimate the upper bound for the number of interactions (steps) executed by an algorithm. Formally, for $f(n) = O(g(n))$ we say that $g(n)$ is an upper bound for $f(n)$, where $f(n)$ is the maximum number of interactions of an algorithm [32]. Note, that *Big-O* refers to the worst-case complexity of the algorithm.
- *Polynomial bound of an algorithm* means that the upper bound of the algorithm can be represented in the form n^c for c greater than 0 [32]. In other words, algorithm for which the number of interactions is estimated as $f(n) = O(n^c)$ for $c > 0$ is bounded by a polynomial in the size of inputs n.
- *Exponential bound of an algorithm* means the bounds in the form c^n for $c > 1$. In other words, algorithm for which the number of interactions is estimated as $f(n) = O(c^n)$ for $c > 1$ is bounded by an exponential in the size of inputs n.
- *Total polynomial time*, *polynomial complexity* or just *polynomial time* means that the total run-time of an algorithm to generate all solutions (outputs) is bounded by a polynomial in the size of the input (cf. [18]).
- *Exponential complexity* or just *exponential time* means that the total run-time of an algorithm to generate all solutions (outputs) is bounded by an exponential in the size of the input (cf. [18]).
- *Polynomial delay* means that an algorithm generates results in such a way, that the time between subsequent outputs is bounded by a polynomial in the size of the input (cf. [12, 18]). We will say, that *subsequent outputs* are generated (computed, calculated) in *polynomial time*.

References

1. Aho AV, Hopcroft JE (1974) The design and analysis of computer algorithms. Addison-Wesley Longman Publishing Co. Inc., Boston
2. Barkaoui K, Minoux M (1992) A polynomial-time graph algorithm to decide liveness of some basic classes of bounded Petri nets. In: Proceedings of the 13th international conference on application and theory of Petri nets, Sheffield, UK, pp 62–75
3. Best E (1987) Structural theory of Petri nets: the free choice hiatus. In: Lecture notes in computer science, vol 254, New York, 1987. Springer, pp 168–206
4. Best E, Thiagarajan P (1987) Some classes of live and safe Petri nets. In: Voss K, Genrich H, Rozenberg G (eds) Concurrency and nets. Springer, Berlin, pp 71–94
5. Blanchard M (1979) Comprendre., Matriser Et Appliquer Le GrafcetCepadues, Toulouse
6. Commoner F, Holt AW, Even S, Pnueli A (1971) Marked directed graphs. J Comput Syst Sci 5(5):511–523
7. David R, Alla H (1992) Petri nets and grafcet: tools for modelling discrete event systems. Prentice-Hall Inc., Upper Saddle River
8. David R, Alla H (2010) Discrete, continuous, and hybrid Petri nets, 2nd edn. Springer, Incorporated
9. Desel J, Esparza J (1995) Free choice Petri nets. Cambridge University Press, New York
10. Desel J, Neuendorf K-P, Radola M-D (1996) Proving nonreachability by modulo-invariants. Theoret Comput Sci 153(1&2):49–64
11. Diaz M (1987) Applying Petri net based models in the design of systems. In: Voss K, Genrich H, Rozenberg G (eds) Concurrency and nets. Springer, Heidelberg, pp 71–94
12. Eiter T (1994) Exact transversal hypergraphs and application to boolean μ-functions. J Symbolic Comput 17(3):215–225
13. Esparza J, Silva M (1990) Top-down synthesis of live and bounded free choice nets. In: Applications and Theory of Petri Nets'90, pp 118–139
14. Hippo homepage. http://www.hippo.iee.uz.zgora.pl. Accessed 4 March 2016
15. Hack M (1974) Analysis of production schemata by Petri nets. MIT Project MAC TR-94 (1972) Corrections: Project MAC. Computation Structures Note 17
16. Holt A, Commoner F (1970) Events and conditions: introduction. In: Dennis JB (ed) Record of the project MAC conference on concurrent systems and parallel computation. ACM, New York, pp 3–5
17. Hu H, Zhou M, Li Z (2012) Liveness and ratio-enforcing supervision of automated manufacturing systems using Petri nets. IEEE Trans Syst Man Cybern Part A Syst Humans 42(2):392–403
18. Johnson DS, Papadimitriou CH, Yannakakis M (1988) On generating all maximal independent sets. Inf Process Lett 27(3):119–123
19. Karatkevich A (2007) Dynamic analysis of Petri net-based discrete systems, vol 356, Lecture notes in control and information sciences. Springer, Berlin
20. Karatkevich A, Wiśniewski R (2015) Relation between SM-covers and SM-decompositions of Petri nets. In: 11th International conference of computational methods in sciences and engineering (ICCMSE), volume 1702 of AIP conference proceedings, Greece, Athens, pp 1–4
21. Kemper P (2004) $O(|P||T|)$-algorithm to compute a cover of S-components in EFC-nets
22. Knuth DE (1997) The art of computer programming: fundamental algorithms, vol 1, 3rd edn. Addison Wesley Longman Publishing Co., Inc, Redwood City, CA, USA
23. Kozlowski T, Dagless E, Saul J, Adamski M, Szajna J (1995) Parallel controller synthesis using Petri nets. IEE Proc Comput Dig Tech 142(4):263–271
24. Lasota A (2012) Modeling of production processes with UML activity diagrams and Petri nets. PhD thesis, University of Zielona Góra
25. Murata T (1989) Petri nets: properties, analysis and applications. Proc IEEE 77:548–580
26. Papadimitriou HC (1994) Computational complexity. Addison Wesley
27. Peterson JL (1981) Petri net theory and the modeling of systems. Prentice Hall PTR, Upper Saddle River

28. Petri CA (1962) Kommunikation mit automaten. Bonn: Institut fr Instrumentelle Mathematik, Schriften des IIM Nr. 2
29. Roguska A (2001) Evaluation of the practical interpreted Petri nets, grafcet networks and networks based on the SEC binary control system design, 2001. Bachelors Thesis, Technical University of Zielona Góra
30. Rozenberg G, Engelfriet J (1998) Elementary net systems. In: Lectures on Petri nets I: basic models, advances in Petri nets, the volumes are based on the Advanced course on Petri nets, London, UK, 1998. Springer, pp 12–121
31. Silva M (2013) Half a century after Carl Adam Petri's Ph.D. thesis: a perspective on the field. Ann Rev Control, 37(2):191–219
32. Sipser M (1996) Introduction to the theory of computation, 1st edn. International Thomson Publishing
33. Valette R (1978) Comparative study of switching representation tool with GRAFCET and Petri nets. Nouv Autom 23(12):377–382
34. Van Der Aalst W, Hee V (2004) Workflow management: models, methods, and systems. The MIT Press
35. Wiśniewska M (2012) Application of hypergraphs in decomposition of discrete systems, vol 23, Lecture notes in control and computer science. University of Zielona Góra Press, Zielona Góra
36. Wiśniewski R, Grobelna I, Grobelny M, Wiśniewska M (2015) Design and verification of distributed logic controllers with application of Petri nets. In: 11th international conference of computational methods in sciences and engineering (ICCMSE), volume 1702 of AIP conference proceedings, Greece, Athens, pp 1–4
37. Wiśniewski R, Stefanowicz Ł, Bukowiec A, Lipiński J (2014) Theoretical aspects of Petri nets decomposition based on invariants and hypergraphs. Lecture notes in electrical engineering, vol 308. Zhangjiajie, China, pp 371–376
38. Wiśniewski R, Grobelna I, Stefanowicz L (2016) Partial reconfiguration of concurrent logic controllers implemented in FPGA devices. In: 12th international conference of computational methods in sciences and engineering—ICCMSE'16, page TBD, Athens, Greece. Accepted for publication
39. Yakovlev A, Gomes L, Lavagno L (2000) Hardware design and Petri nets. Springer
40. Zakrevskij A, Pottosin Y, Cheremisinova L (2009) Design of logical control devices. TUT Press, Moskov

Chapter 3
Perfect Graphs and Comparability Graphs

3.1 Preliminaries

Definition 3.1 (*Graph*) A *graph* or *undirected graph* is defined by a pair

$$G = (V, E), \tag{3.1}$$

where

$V = \{v_1, \dots, v_n\}$, is a finite, nonempty set of vertices;
$E = \{E_1, \dots, E_m\}$, is a finite set of unordered pair of vertices, called edges.

We assume that a *graph* has no loops or multiple edges. For further purposes, an undirected edge incident with vertex u and vertex v is denoted by $\{u, v\}$ (or occasionally just uv). We say that two vertices are *adjacent* (*neighbors*) if they are joined by an edge.

Please note, that simple notation *graphs* we use in reference to *undirected graphs*, in which the set of edges consists of *unordered pairs of vertices*. On contrary, *directed graphs* are composed of *ordered pairs of vertices* are called *digraphs* [13, 24], introduced later in this chapter.

Definition 3.2 (*Vertex degree*) A degree of a vertex v_i denoted by $deg(v_i)$ is the number of edges incident with v_i. The minimum degree among all vertices is denoted by $\delta(G)$, while $\Delta(G)$ is the largest such number [13].

An exemplary undirected graph G_1 is shown in Fig. 3.1 (left). The presented graph consists of $|V| = 6$ vertices: $V = \{v_1, \dots, v_6\}$ and $|E| = 7$ edges: $E = \{E_1, \dots, E_7\}$. Each edge forms a set of unordered pairs of vertices: $E_1 = \{v_1, v_2\}$, $E_2 = \{v_1, v_6\}$, $E_3 = \{v_2, v_3\}$, $E_4 = \{v_2, v_4\}$, $E_5 = \{v_2, v_5\}$, $E_6 = \{v_3, v_4\}$ and $E_7 = \{v_4, v_5\}$.

© Springer International Publishing Switzerland 2017

R. Wiśniewski, *Prototyping of Concurrent Control Systems Implemented in FPGA Devices*, Advances in Industrial Control,
DOI 10.1007/978-3-319-45811-3_3

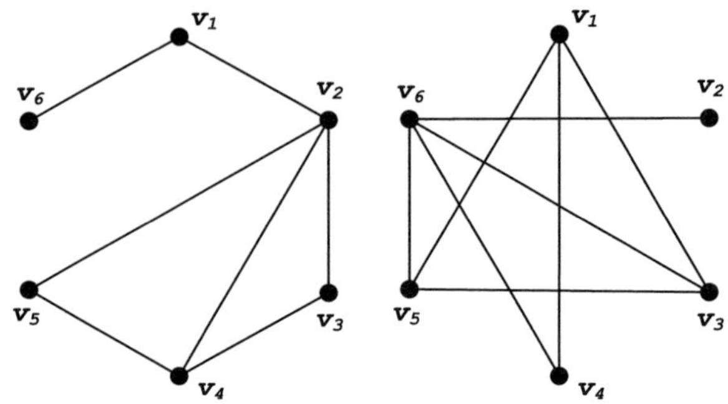

Fig. 3.1 Exemplary graph G_1 (*left*) and its complementary graph $\overline{G_1}$ (*right*)

The minimum degree of G_1 is equal to $\delta(G) = deg(v_6) = 1$, while its maximum degree $\Delta(G) = deg(v_2) = deg(v_4) = 3$.

Definition 3.3 (*Complement graph*) A *complement graph* $\overline{G} = (V, \overline{E})$ of a graph $G = (V, E)$ is a graph with the same set of vertices, and the set of edges $\overline{E} = \{(u, v) \in V \times V \mid u \neq v \text{ and } (u, v) \notin E\}$. In other words, two vertices of $\overline{G}$ are connected by an edge if they are not adjacent in G.

The complement graph $\overline{G_1}$ of a graph G_1 is shown in Fig. 3.1 (right). The set of vertices remains unchanged, thus $|V| = 7$. There are totally $|\overline{E}| = 8$ edges in $\overline{G_1}$. Particular edges connect vertices that are not neighbors in G_1.

The minimum degree of G_1 is equal to $\delta(G) = deg(v_2) = 1$, while its maximum degree $\Delta(G) = deg(v_6) = 4$.

Definition 3.4 (*Subgraph*) A *subgraph* $G' = (V', E')$ of graph G is a graph, where $V' \subseteq V$ and $E' \subseteq E$ [3].

Definition 3.5 (*Induced subgraph*) A graph G' is called an *induced subgraph* of G if for any pair of vertices u and v of G', (u, v) is an edge of G' if and only if (u, v) is an edge of G.

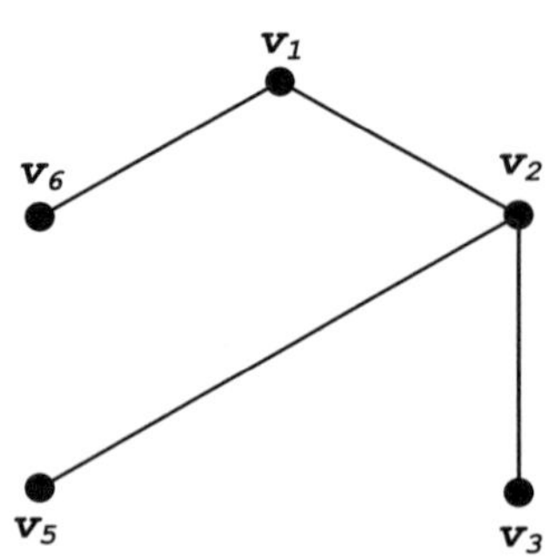

Fig. 3.2 Induced subgraph $\overline{G'_1}$ of a graph G_1

Figure 3.2 shows an exemplary induced subgraph $G'_1 = (V', E')$ of G_1. The set of vertices contains $|V'| = |V|\backslash\{v_4\} = 5$ vertices, while the set of edges is reduced to $|E'| = |E|\backslash\{E_4, E_6, E_7\} = 4$ edges.

One of the graph representations is *an incidence matrix*. The rows of the matrix refer to edges, while the columns are related to graph vertices. If an incidence matrix element equals 1, the jth edge ($j \in 1, \ldots, m$) is incident to the ith vertex ($i \in 1, \ldots, n$). Otherwise, the element equals 0

$$M = \begin{cases} 1 & \text{if} \quad v_i \in E_j \\ 0 & \text{if} \quad v_i \notin E_j \end{cases}, \tag{3.2}$$

Incidence matrix M_{G_1} of graph G_1 is shown in Fig. 3.3. Graph G_1 consists of six vertices, represented by matrix columns. Similarly, seven edges are represented by rows of the matrix.

Another common representation of a graph is a *neighborhood matrix* (also called *an adjacency matrix*). It is a symmetric matrix including relations between particular vertices. The ith row and jth column of the matrix determine the number of edges joining the ith and the jth vertices.

Neighborhood matrix N_{G_1} for graph G_1 is presented in Fig. 3.4.

Definition 3.6 (*Clique*) A *clique* of graph G is a subset of vertices $Q \subseteq V$, such that any two vertices of Q are connected by an edge. A *maximal clique* is a clique that cannot be extended by any other vertex. A *clique number* of G is equal to the maximum cardinality of the clique in G and is denoted by $\omega(G)$. *Clique cover* is the set of cliques that cover all the vertices of a graph. A *clique cover number* is the smallest number of cliques that are needed to cover all the vertices and is denoted by $\theta(G)$ [11].

Fig. 3.3 Incidence matrix M_{G_1} of graph G_1

$$M_{G_1} = \begin{array}{c} \begin{array}{cccccc} v_1 & v_2 & v_3 & v_4 & v_5 & v_6 \end{array} \\ \begin{bmatrix} 1 & 1 & 0 & 0 & 0 & 0 \\ 1 & 0 & 0 & 0 & 0 & 1 \\ 0 & 1 & 1 & 0 & 0 & 0 \\ 0 & 1 & 0 & 1 & 0 & 0 \\ 0 & 1 & 0 & 0 & 1 & 0 \\ 0 & 0 & 1 & 1 & 0 & 0 \\ 0 & 0 & 0 & 1 & 1 & 0 \end{bmatrix} \end{array} \begin{array}{c} \\ E_1 \\ E_2 \\ E_3 \\ E_4 \\ E_5 \\ E_6 \\ E_7 \end{array}$$

Fig. 3.4 Neighborhood (adjacency) matrix of graph G_1

$$N_{G_1} = \begin{array}{c} \begin{array}{cccccc} v_1 & v_2 & v_3 & v_4 & v_5 & v_6 \end{array} \\ \begin{bmatrix} 0 & 1 & 0 & 0 & 0 & 1 \\ 1 & 0 & 1 & 1 & 1 & 0 \\ 0 & 1 & 0 & 1 & 0 & 0 \\ 0 & 1 & 1 & 0 & 1 & 0 \\ 0 & 1 & 0 & 1 & 0 & 0 \\ 1 & 0 & 0 & 0 & 0 & 0 \end{bmatrix} \end{array} \begin{array}{c} \\ v_1 \\ v_2 \\ v_3 \\ v_4 \\ v_5 \\ v_6 \end{array}$$

Fig. 3.5 Cliques of a graph G_1

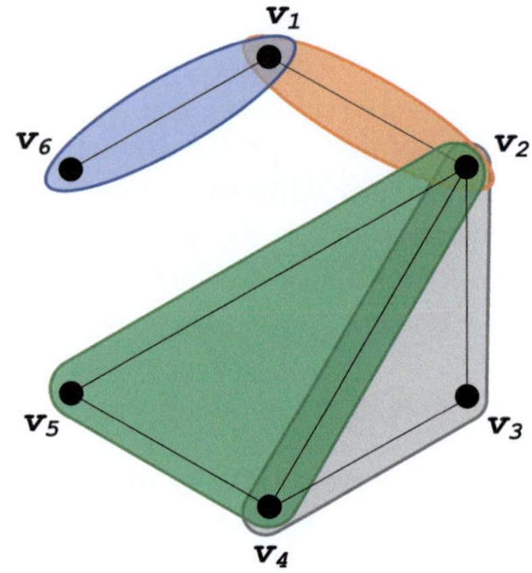

There are totally four cliques in graph G_1. The clique number of G_1 is equal to $\omega(G_1) = 3$. There are two cliques with the maximum cardinality: $Q_1 = \{v_2, v_3, v_4\}$ and $Q_2 = \{v_2, v_4, v_5\}$. Each of the remaining two cliques contain two vertices: $Q_3 = \{v_1, v_2\}$, $Q_4 = \{v_1, v_6\}$. Figure 3.5 illustrates the cliques of G_1. Notice, that vertices $\{v_2\}$ and $\{v_4\}$ are shared between two cliques: Q_1 and Q_2. The clique cover number of this graph is equal to $\theta(G_1) = 3$, since it is the smallest number of cliques to cover all the vertices. Such a cover can be achieved by cliques $\{Q_1, Q_2, Q_4\}$.

Definition 3.7 (*Compatibility relation of vertices*) Two or more vertices of a graph G are *compatible* (or in *compatibility relation*) if they are not incident to the same edge. In other words: two or more vertices are in *compatibility relation* if they are not connected by any edge.

Definition 3.8 (*Independent set, stable set*) A set $I \subseteq V$ is an *independent set* (also called *stable set*) of a graph G if all vertices that belong to I are compatible. Set I is called maximal if no $I' \supset I$ is an independent set of G [16]. A *stability number* of a graph is the number of vertices in an independent set of maximum cardinality and is denoted by $\alpha(G)$ [10].

There are five maximal independent sets of G_1: $I_1 = \{v_1, v_4\}$, $I_2 = \{v_1, v_3, v_5\}$, $I_3 = \{v_2, v_6\}$, $I_4 = \{v_3, v_5, v_6\}$, and $I_5 = \{v_4, v_6\}$. The stability number of G_1 is $\alpha(G) = 3$.

Obtaining all independent sets in a graph, as well as generating its all cliques are well-known combinatoric problems [6, 16, 21, 23]. Furthermore, such problems are complementary, since an independent set refers to a clique in a graph complement and vice versa. Generally, both problems are classified as NP-complete, which means, that there is no known polynomial time algorithm to solve the task [1, 16, 23].

The basic algorithm for computation of all cliques in a graph simply uses *backtracking* concept (cf. [1]), like classic *Bron–Kerbosh* method [4]. The particular cliques are obtained by recursively adding adjacent vertices to the set and checking

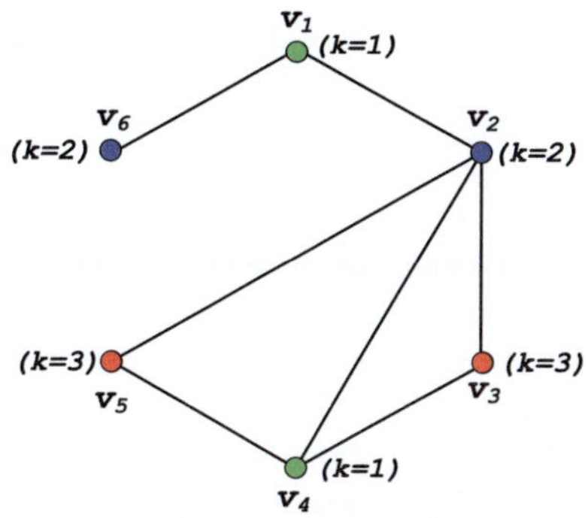

Fig. 3.6 A possible coloring of graph G_1

if the set still forms a clique. The extension of such a method uses a technique called *pivoting* that avoids searching for non-maximal cliques [23]. Another enhancement of the classical *Bron–Kerbosh* method uses *vertex ordering* to reduce the recursive calls of algorithm (cf. [4, 23]).

Definition 3.9 (*Graph coloring*) A *graph coloring* is the assignment of one selected color to each vertex such that no two adjacent vertices share the same color [8]. A *chromatic number* of a graph is equal to the smallest number of colors by which the graph can be colored and is denoted by $\chi(G)$.

Coloring is closely related to independent sets of a graph. In fact, coloring splits the graph vertices into independent sets. However, particular vertices may occur in different independent sets, which are not allowed in the case of classical graph coloring, where only one color is assigned to the vertex.

The smallest number of colors by which G_1 can be colored is equal to $\chi(G) = 3$. Figure 3.6 shows one of the possible coloring of G_1 (particular colors are denoted by a variable k):

- first color ($k = 1$): $\{v_1, v_4\}$,
- second color ($k = 2$): $\{v_2, v_6\}$,
- third color ($k = 3$): $\{v_3, v_6\}$.

Note, that there exist other versions of coloring of G_1. For example, alternate assignments of colors of G_1 may result in the following coloring:

- first color ($k = 1$): $\{v_1, v_3, v_5\}$,
- second color ($k = 2$): $\{v_2\}$,
- third color ($k = 3$): $\{v_4, v_6\}$.

Obtaining the chromatic number of a graph as well as optimal coloring (with the smallest number of colors) are NP-complete problems. Therefore, existing methods balance between optimal results (exact algorithms) and reasonable time (approximate solutions) [1, 13, 15, 18, 24]. Both groups have advantages and weak points.

Exact methods, like classical *sequential backtracking algorithm* (cf. [1]), search recursively the whole set of results to find the best solution. Unfortunately, they run in exponential time, which means that the solution may be never found.

Approximate graph coloring algorithms usually base on the greedy algorithm [15, 18]. The methods search for the local optimum to find the best solution, however they do not guarantee optimal results. The most popular approximate graph coloring methods apply heuristic vertex ordering, like *Largest-first* or *Smallest-last*. Such ideas may increase the chance for finding the best solution, but they still do not assure it [15].

From the above we see, that in the general case, the problem of graph coloring requires a compromise between optimal results and computational time. However, there exists a special group of graphs called *perfect graphs*, where such a problem can be solved polynomially.

3.2 Perfect Graphs

This section deals with a special class of graphs. A brief overview and examples of perfect graphs are given. *Perfect graphs* have unique properties. For example, the recognition of a perfect graph, obtaining its clique number or its chromatic number as well as the graph coloring are computable in polynomial time [7, 10, 20, 22].

Definition 3.10 (*Perfect graph*) A graph G is *perfect* if, for every induced subgraph G' of G, $\chi(G') = \omega(G')$ [7].

Note, that the above definition implies another property: for every induced subgraph G' of G, the condition $\theta(G') = \alpha(G')$ also ought to be fulfilled [19].

Recall graph G_1 shown in Fig. 3.1 (left). As it was already shown, a chromatic number of this graph is equal to its clique number $\chi(G) = \omega(G) = 3$. Moreover, its clique cover number is equal to the stability number: $\theta(G) = \alpha(G) = 3$. Furthermore, every induced subgraph of G_1 holds these properties, which means that G_1 is perfect.

Let us now analyze a complementary graph $\overline{G_1}$ of G_1, shown in Fig. 3.1 (right). From the earlier analysis, we know that there are $|I| = 5$ maximal independent sets in G_1. Since each maximal independent set of a graph refers to a clique in its complement, we immediately know that there are $|Q| = 5$ cliques in $\overline{G_1}$. All the cliques of $\overline{G_1}$ are presented in Fig. 3.7 (left). A clique number of this graph is equal to the stability number of G_1: $\omega(\overline{G_1}) = \alpha(G_1) = 3$. The smallest number of colors by which $\overline{G_1}$ can be colored is $\chi(\overline{G_1}) = 3$, as shown in Fig. 3.7 (right). It means, that its chromatic number is the same as clique number: $\chi(\overline{G_1}) = 3\chi(\overline{G_1})$. Furthermore, every induced subgraph $\overline{G_1'}$ of $\overline{G_1}$ also satisfies this property. Therefore, both graphs G_1 and its complement $\overline{G_1}$ are perfect.

In fact, there is a theorem that proves our analysis:

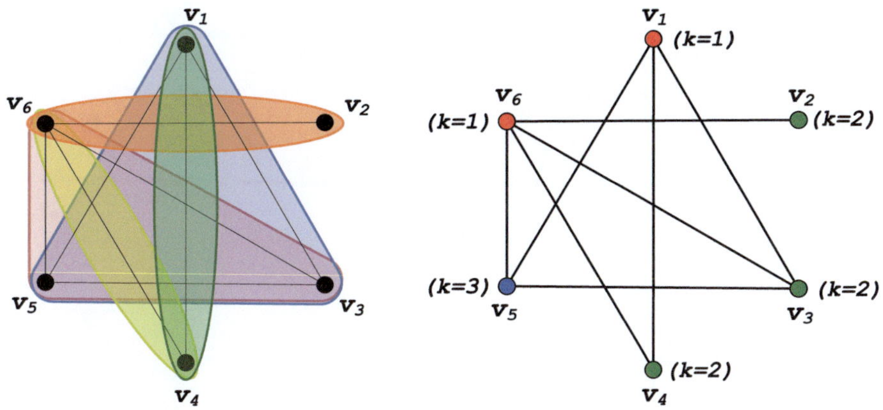

Fig. 3.7 Cliques (*left*) and possible coloring (*right*) of graph $\overline{G_1}$

Fig. 3.8 Graph G_2

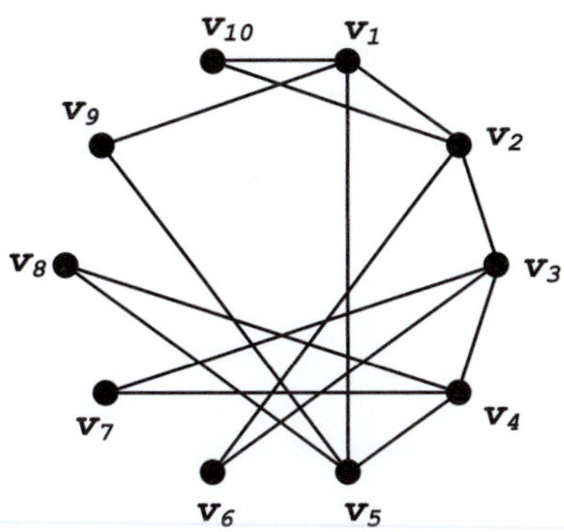

Theorem 3.1 *The complement of a perfect graph is perfect as well (cf. [19]).*

Figure 3.8 shows another graph. G_2 contains $|V| = 10$ vertices, connected by $|E| = 15$ edges. The chromatic number of G_2 is equal to $\chi(G_2) = 3$, as it is presented in Fig. 3.9 (left). Colors split the graph vertices into three maximal independent sets: $I_1 = \{v_1, v_3, v_8\}$, $I_2 = \{v_2, v_4, v_9\}$, $I_3 = \{v_5, v_6, v_7\}$. Furthermore, there are totally $|Q| = 5$ cliques of G_2: $Q_1 = \{v_1, v_2, v_{10}\}$, $Q_2 = \{v_2, v_3, v_6\}$, $Q_3 = \{v_3, v_4, v_7\}$, $Q_4 = \{v_4, v_5, v_8\}$, $Q_2 = \{v_1, v_5, v_9\}$. The cliques of G_2 are shown in Fig. 3.9 (right). The cardinality of all the cliques is the same and the clique number of G_2 is equal to $\omega(G_2) = 3$. It means, that the condition $\chi(G_2) = \omega(G_2)$ is satisfied.

Let us now analyze an induced subgraph of G_2. Figure 3.10 (left) illustrates induced subgraph G_2' of initial graph G_2 by the set of vertices $\{v_6, v_7, v_8, v_9, v_{10}\}$. According to the perfect graph theorem, the chromatic number of this subgraph should be equal to its clique number. Subgraph G_2' can be optimally colored with the use of $\chi(G_2') = 3$ colors, as it is presented in Fig. 3.10 (right). On the other hand, the cardinality of all the cliques of G_2' is equal to $\omega(G_2') = 2$. Clearly, each edge of the graph forms a maximal clique, that cannot be extended. It means, that for the induced

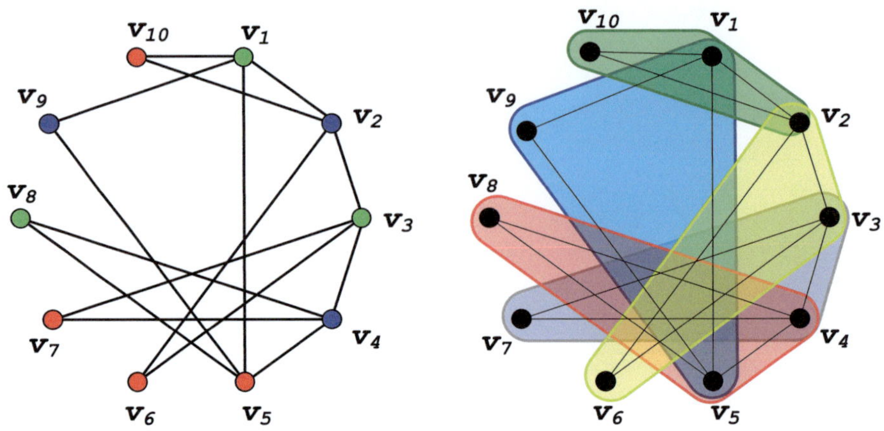

Fig. 3.9 Coloring (*left*) and cliques (*right*) of graph G_2

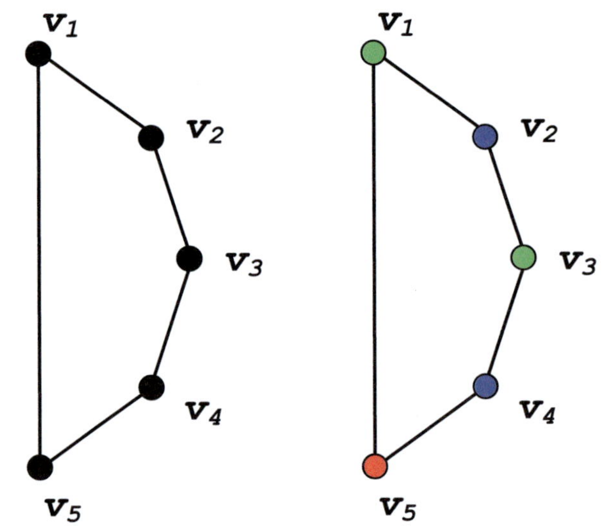

Fig. 3.10 Induced subgraph G_2' of G_2 (*left*) and its coloring (*right*)

subgraph G_2', the chromatic number is not equal to the clique number. Therefore, graph G_2 is not perfect, and hence neither its complement.

3.3 Comparability Graphs

The class of *comparability graphs* is the subclass of *perfect graphs*. It means, that comparability graphs have all the properties of perfect graphs, and in addition, they reveal some unique ones.

Before we formally define a comparability graph, let us introduce auxiliary notations and definitions [2, 5, 10, 13, 14, 17, 22, 24].

Definition 3.11 (*Digraph*) A *digraph* (*directed graph*) is defined by a pair:

$$D = (V, F), \tag{3.3}$$

where:

$V = \{v_1, \dots, v_n\}$, is a finite, nonempty set of vertices;
$F = \{F_1, \dots, F_m\}$, is a finite set of ordered pair of vertices, called edges.

By a definition, a digraph has no loops or multiple arcs [13]. For further purposes, directed edges are denoted by (u, v) or just $\overrightarrow{uv}$ (to simplify enumerations of edges), where u is the first vertex (tail), and v is the second vertex (head) of an edge [2, 5].

Definition 3.12 (*Path*) A *path* in a digraph is a sequence of vertices $[v_1, \dots, v_l]$ of *length* l such that for every $i \in (1, \dots, l)$ there is an edge $(v_{i-1}, v_i) \in F$ [10]. A path is *simple* if each vertex occurs at least once [10]. A path is *trivial* if $l = 0$.

Definition 3.13 (*Cycle*) A *cycle* in a digraph is a sequence of vertices $[v_1, \dots, v_l]$ such that for every $i \in (1, \dots, l)$ there is an edge (v_{i-1}, v_i), and additionally $(v_i, v_1) \in F$ [10, 14]. A digraph is *acyclic* if it does not contain any cycles.

Definition 3.14 (*A transitive digraph*) A digraph is *transitive* if, whenever (u, v) and (v, z) are directed edges, (u, z) is also a directed edge [12].

Definition 3.15 (*A reversal digraph*) Let $D = (V, E)$ be a digraph, then digraph $D^{-1} = (V, E^{-1})$ is *reversal* of D, where $E^{-1} = \{(u, v) \mid (v, u) \in E\}$.

Definition 3.16 (*Topological ordering*) A *topological ordering* of a digraph $D = (V, F)$ is a linear ordering of the vertices $[v_1, v_2, \dots, v_n]$ which is consistent with F, that is

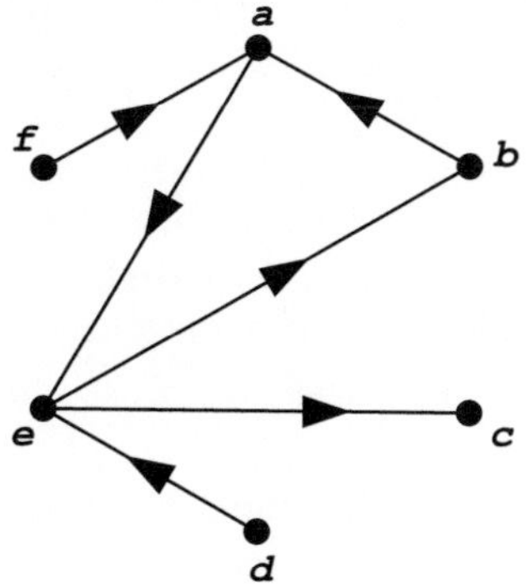

Fig. 3.11 Exemplary directed graph D_1

$$\forall (i, j \in n) : \overrightarrow{v_i v_j} \in F \implies i < j \tag{3.4}$$

An ordering which fulfills (3.4) is called a *topological sorting* of D [10].

An exemplary digraph D_1 is shown in Fig. 3.11. It consists of six vertices $V = \{a, b, c, d, e, f\}$ denoted by subsequent letters of alphabet, and six directed edges $E = \{\overrightarrow{ba}, \overrightarrow{ae}, \overrightarrow{fa}, \overrightarrow{eb}, \overrightarrow{ec}, \overrightarrow{de}\}$. A sequence of vertices $[a, e, b]$ forms a cycle in D_1, since there exist edges: $\overrightarrow{ae}, \overrightarrow{eb}, \overrightarrow{ba}$. An exemplary simple path may be formed by a sequence of vertices $[f, a, e, c]$.

Definition 3.17 (*Comparability graph*) An undirected graph $G = (V, E)$ is a *comparability graph* (also called a *transitively orientable* (*TRO*) *graph*) if there exists an orientation (V, F) of G that satisfies the following condition:

$$\begin{aligned} F \cap F^{-1} &= \emptyset, \\ F \cup F^{-1} &= E, \\ F^2 &\subseteq F, \end{aligned} \tag{3.5}$$

where $F^2 = \{(u, z) \mid (u, v)\,,\ (v, z) \in F \text{ for some vertex } v\}$.

The relation F is a strict partial ordering of V comparability relation of which is E, and then F is called a *transitive orientation* of G [10].

In other words, a graph is a comparability graph if its edges can be oriented in such a way, that the resulting digraph is *transitive* and *acyclic* [12].

Definition 3.18 (*Binary orientation* Γ) A binary orientation Γ of an undirected graph $G = (V, E)$ is defined as follows [10]:

$$(u, v)\ \Gamma\ (u', v') \Leftrightarrow \begin{cases} \text{either } u = u' \text{ and } v\,v' \notin E \\ \text{or} \quad\ \ v = v' \text{ and } u\,u' \notin E. \end{cases} \tag{3.6}$$

It is said, that (u, v) directly forces (u', v') whenever $(u, v)\ \Gamma\ (u', v')$ [22].

Definition 3.19 (*Implication classes*) Let Γ^* be the reflexive, transitive closure of Γ (cf. [10]). Then, Γ^* is an equivalence relation on E and partitions E into the *implication classes* of G. Furthermore, edges (u, v) and (y, z) are in the same *implication class* if there exists a sequence of edges [10]:

$$(u, v) = (u_0, v_0)\ \Gamma\ (u_1, v_1)\ \Gamma\ \cdots\ \Gamma\ (u_k, v_k) = (y, z), \quad \text{with } k \geq 0. \tag{3.7}$$

Graph G_3 shown in Fig. 3.12 (left) is a comparability graph. It can be transitively oriented, as it is illustrated in Fig. 3.12 (right). The orientation of particular edges is as follows: $\overrightarrow{ea}, \overrightarrow{ab}, \overrightarrow{eb}, \overrightarrow{cb}, \overrightarrow{cd}, \overrightarrow{ed}$. Note, that reverse orientation of edges also results in a TRO graph: $\overrightarrow{ae}, \overrightarrow{ba}, \overrightarrow{be}, \overrightarrow{bc}, \overrightarrow{dc}, \overrightarrow{de}$.

Let us now analyze graph G_4 presented in Fig. 3.13 (left). An attempt of transitive orientation is shown in Fig. 3.12 (right). Starting from the edge $\overrightarrow{ab}$, particular edges

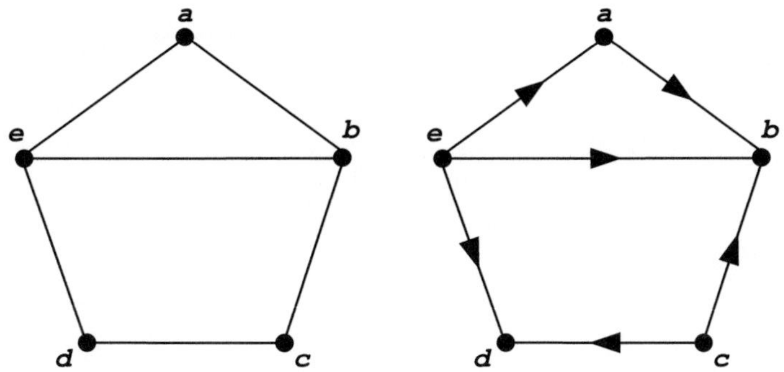

Fig. 3.12 Graph G_3 (*left*) and its transitive orientation (*right*)

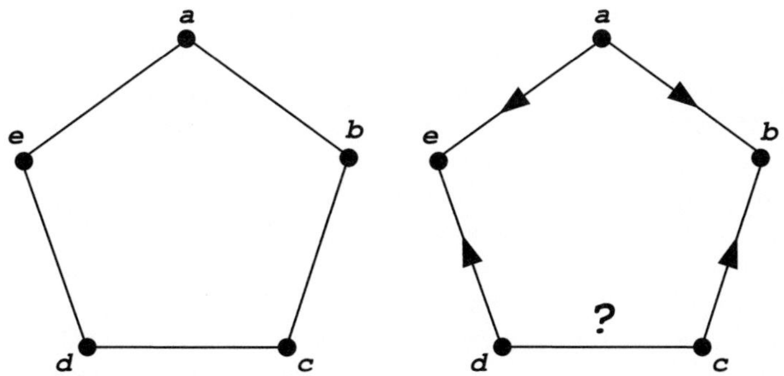

Fig. 3.13 Graph G_4 (*left*) and attempt of its orientation (*right*)

are oriented as follows: $\overrightarrow{ae}$, $\overrightarrow{bc}$ (directly forced by $\overrightarrow{ab}$), $\overrightarrow{de}$ (forced by $\overrightarrow{ae}$). There is no possibility to make an orientation of the remaining edge $\{c, d\}$. Orientation $\overrightarrow{cd}$ is forced by $\overrightarrow{cb}$, while the reverse orientation $\overrightarrow{dc}$ is forced by $\overrightarrow{de}$. On the other hand, both orientations violate the condition (3.5). Sequence $\overrightarrow{dc}$, $\overrightarrow{cb}$ requires an orientation of a missing edge $\overrightarrow{db}$. Similarly, orientations $\overrightarrow{cd}$ and $\overrightarrow{de}$ forces an orientation $\overrightarrow{ce}$ of non-existing edge. Therefore, graph G_4 is not a comparability graph.

3.4 Recognition and Coloring of Comparability Graphs

In this section, the ideas of comparability graph recognition and coloring are shown. Most popular methods are briefly described. Further, we introduce a novel technique for recognition of comparability graphs. Finally, we propose a combined solution that enables the computation of TRO and the assignment of colors to particular vertices simultaneously.

3.4.1 *Recognition of Comparability Graphs*

According to the definition, a graph is a comparability graph if it can be transitively oriented. Existing algorithms strictly rely on this property. Particularly, implication classes are enumerated. A graph is not a comparability graph, if there is an attempt to force a direction $\overrightarrow{vu}$ of an already oriented edge $\overrightarrow{uv}$ in a particular implication class [10]. Most known (classical) recognition methods apply a *depth-first search* (*DFS*) algorithm (cf. [1]). The idea bases on the decomposition of the graph vertices into implication classes, by a recursive orientation of subsequent edges [9, 10, 22]. At the beginning, an initial edge is selected arbitrary. Furthermore, an implication class B_i containing such an edge is enumerated, as well as a class containing reverse ordering $B_i{}^{-1}$. According to (3.5), a graph can be transitively oriented if $F \cap F^{-1} = \emptyset$. Since $B_i \in F$, the condition $B_i \cap B_i^{-1} = \emptyset$ also must be satisfied. If the property is violated, the graph is not a comparability graph [10].

Another interesting approach was shown in [12], where a lexicographic *breadth-first search* (*BFS*) search algorithm is used (cf. [1, 12]). The method also decomposes the set of vertices into partition classes. Additionally, an operation called *pivoting* (cf. [12]) is used to split vertices into particular partition classes according to their adjacency.

Let us introduce an alternate version of comparability graph recognition. Algorithm 3.1 computes the transitive orientation F of an undirected graph $G = (V, E)$. The algorithm explores subsequent edges with the use of a modified *BFS* method. Subsequent transitive orientations are added to set F, while set $\mathcal{E}$ (a copy of E) holds unexplored edges.

Initially, edge $\{u, v\}$ is selected arbitrary. Its orientation $\overrightarrow{uv}$ is added to the queue Q. In the further steps, the inner *while* loop is executed. First, edge $\overrightarrow{uv}$ is dequeued from Q. The existence of a reverse orientation $\overrightarrow{vu}$ in the set F means that there is an attempt to force already oriented edge and G is not a comparability graph. Otherwise, $\overrightarrow{vu}$ is added to F, while $\{u, v\}$ is removed from set $\mathcal{E}$. Afterwards, edges directly forced by $\overrightarrow{uv}$ are explored. For each vertex z that is adjacent to u or v (but their common neighbors, except directly forced edges by $\overrightarrow{vz}$ or $\overrightarrow{zu}$), an oriented edge $\overrightarrow{uz}$ (or $\overrightarrow{zv}$, respectively) is selected. If such an orientation has not been already added to the queue nor to set F, it is enqueued to Q. The inner *while* loop is executed until

Algorithm 3.1 Computation of a TRO of an undirected graph

Input: An undirected graph $G = (V, E)$
Output: A transitive orientation F of G

```
ℰ ← E
F ← ∅
Q.clear()
do
    select arbitrary edge {u, v} ∈ ℰ
    Q.push (uv→)
    while Q is not empty do
        uv→ = Q.pop()
        if vu→ ∈ F then
            Output 'Not a comparability graph'
            return ∅
        end if
        F = F ∪ uv→
        ℰ = ℰ \ {u, v}
        for each z=Adj(u) such that [z≠Adj(v) or vz→ ∈ F] do
            if uz→ ∉ {F ∪ Q} then Q.push (uz→)
        end for
        for each z=Adj(v) such that [z≠Adj(u) or zu→ ∈ F] do
            if zv→ ∉ {F ∪ Q} then Q.push (zv→)
        end for
    end while
while ℰ ≠ ∅
return F
```

the queue is empty. The outer *do...while* loop repeats the above procedure for all unexplored edges of G, that is, while there are remaining edges in $\mathcal{E}$.

Note, that the initial orientation of edge $\{u, v\}$ is essential for the whole transitive orientation. There is a possibility to reverse the orientation ($\overrightarrow{vu}$, instead of $\overrightarrow{uv}$). Such a modification results in a reverse orientation of all remaining edges, which are forced by $\overrightarrow{vu}$.

In opposition to the already known TRO algorithms, the presented solution does not enumerate implication classes, nor partitions vertices into subsets.

Let us illustrate the above algorithm by an example. From the previous analysis we know, that graph G_1 is perfect. Now we check, whether it is a comparability graph. At the beginning $\mathcal{E} = E = \{\{v_1, v_2\}, \{v_1, v_6\}, \{v_2, v_3\}, \{v_2, v_4\}, \{v_2, v_5\}, \{v_3, v_4\}, \{v_4, v_5\}\}$, $F = \emptyset$, $Q = \emptyset$.

Let us start the procedure by selecting of $\{v_1, v_2\}$. Orientation $\overrightarrow{v_1v_2}$ of such an edge is added to queue Q. In the inner *do...while* loop $\overrightarrow{v_1v_2}$ is dequeued and added to set F, while $\{v_1, v_2\}$ is removed from $\mathcal{E}$. At the first *for each* loop, edge $\overrightarrow{v_1v_6}$ is added to Q, since $z = v_6$ is the only neighbor of v_1. Execution of the second *for each* loop results in a queue of three edges: $\overrightarrow{v_3v_2}$, $\overrightarrow{v_4v_2}$, $\overrightarrow{v_5v_2}$. At this point $\mathcal{E} = \{\{v_1, v_6\}, \{v_2, v_3\}, \{v_2, v_4\}, \{v_2, v_5\}, \{v_3, v_4\}, \{v_4, v_5\}\}$, $F = \{\overrightarrow{v_1v_2}\}$, $Q = \{\overrightarrow{v_1v_6}, \overrightarrow{v_3v_2}, \overrightarrow{v_4v_2}, \overrightarrow{v_5v_2}\}$.

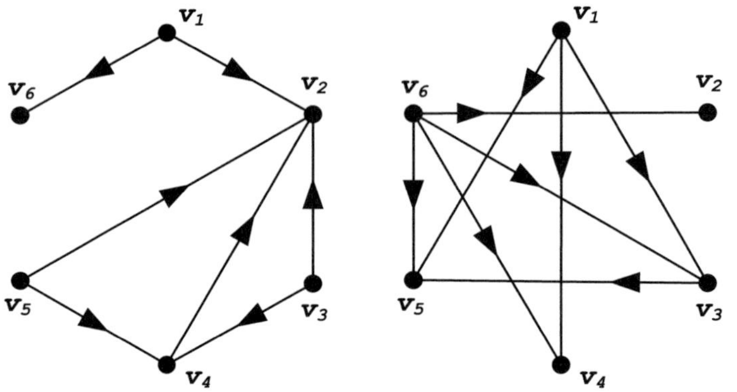

Fig. 3.14 Transitive orientations of G_1 (*left*) and $\overline{G_1}$ (*right*)

At the second execution of the inner *do...while* loop, edge $\{v_1, v_6\}$ is explored. Since F does not contain $\overrightarrow{v_6v_1}$, orientation $\overrightarrow{v_1v_6}$ can be successfully added to F, and $\{v_1, v_6\}$ is removed from $\mathcal{E}$. At this stage no edges are queued, since edge $\{v_1, v_2\}$ has been already explored and v_6 does not have any neighbors but v_1. Similarly, the exploration of the three further edges results in adding $\overrightarrow{v_3v_2}$, $\overrightarrow{v_4v_2}$, $\overrightarrow{v_5v_2}$ to set F and removing $\{v_2, v_3\}$, $\{v_2, v_4\}$, $\{v_2, v_5\}$ from $\mathcal{E}$, but no more edges are enqueued. Thus, the first execution of the outer *do...while* loop is finished. At this stage $\mathcal{E} = \{\{v_3, v_4\}, \{v_4, v_5\}\}$, $F = \{\overrightarrow{v_1v_2}, \overrightarrow{v_1v_6}, \overrightarrow{v_3v_2}, \overrightarrow{v_4v_2}, \overrightarrow{v_5v_2}\}$, $Q = \emptyset$.

Further execution of the algorithm results in exploration of the remaining edges of $\mathcal{E}$. Notice, that edge $\{v_3, v_4\}$ can be oriented in any way. Let us make an orientation $\overrightarrow{v_3v_4}$, which directly forces the remaining unexplored edge $\{v_4, v_5\}$ by the second *for each* loop to the orientation $\overrightarrow{v_5v_4}$.

Finally, the algorithm finishes exploration of all edges with a successful orientation of G_1. The transitive orientation $F = \{\overrightarrow{v_1v_2}, \overrightarrow{v_1v_6}, \overrightarrow{v_3v_2}, \overrightarrow{v_4v_2}, \overrightarrow{v_5v_2}, \overrightarrow{v_3v_4}, \overrightarrow{v_5v_4}\}$ is shown in Fig. 3.14 (left).

A complementary graph $\overline{G_1}$ of G_1 is also transitively orientable: $F = \{\overrightarrow{v_1v_3}, \overrightarrow{v_1v_4}, \overrightarrow{v_6v_3}, \overrightarrow{v_1v_5}, \overrightarrow{v_6v_4}, \overrightarrow{v_6v_2}, \overrightarrow{v_6v_5}, \overrightarrow{v_3v_5}\}$, as illustrated in Fig. 3.14 (right).

Let us now analyze the computational complexity of the presented algorithm. Clearly, both *while* loops (outer *do...while* and inner *while*) are executed $|E|$ times, once for each edge of the graph. Furthermore, each of the *for each* loops is executed for all neighbors of a vertex, thus its complexity is at most $O(\Delta(G))$, where $\Delta(G)$ is the maximum degree of vertices in G. Thus, the time complexity of the whole algorithm is bounded by $O(|E| * \Delta(G))$.

Theorem 3.2 *Comparability graph recognition with application of the Algorithm 3.1 is bounded by* $O(|E| * \Delta(G))$*, where* $\Delta(G)$ *is the maximum degree of a vertex.*

3.4.2 Coloring of Comparability Graphs

Let $D = (V, F)$ be a directed and acyclic digraph. A strict partial ordering of vertices, namely $u > v$, can be associated to the orientation F if and only if there exists a nontrivial path in F from u to v. Furthermore, a *height function* h can be defined on V in such a way:

$$h(v) = \begin{cases} 0 \text{ if } v \text{ is a } sink \text{ (that is, if } \forall u = Adj(v) : \overrightarrow{vu} \notin F), \\ 1 + max\{h(u) \mid \overrightarrow{vu} \in F\} \text{ otherwise.} \end{cases} \quad (3.8)$$

Function h returns a proper (and optimal) vertex coloring of $G = (V, F)$ (cf. [10, 22]). Thus, height $h(v)$ assigns a color to vertex v. In other words, a color of a vertex in a comparability graph is strictly associated to its height.

Let us now extend Algorithm 3.1 by adding a computation of vertices height. Algorithm 3.2 searches for a transitive orientation F and simultaneously calculates height h of vertices of an undirected graph $G = (V, E)$. It is considered, that h is designed as a vector of heights of all vertices. Particular vertices are enumerated in a lexicographic order, that is, $h[1]$ refers to the height of vertex a (or v_1), $h[2]$ means the height of b (or v_2), and so on. Moreover, set T holds a reverse topological ordering of vertices V, that is, $\overrightarrow{uv} \in F \implies T(u) > T(v)$. It is updated each time an oriented edge is added to F. Such an ordering is essential for the computational complexity of the whole algorithm, which we shall show later.

Essential function *IncHeight* is executed each time an oriented edge $\overrightarrow{uv}$ is added to F. Such a function is called only if a height of u is not greater than height of v. If so, the height of u is increased and the function is recursively executed for each neighbor z of u such that $\overrightarrow{zu} \in F$.

A pseudo-code of *IncHeight* is shown as Algorithm 3.3. Before calculation of the computational complexity of function *IncHeight*, let us introduce axillary lemmas. At the beginning we shall prove that the existence of partial ordering between any three vertices is strict at any stage of execution of *Algorithm 3.2* for any graph, not only a comparability graph. Further, the number of recursive calls of an algorithm *Algorithm 3.3* is estimated.

Lemma 3.1 *Let $\{u, v, z\}$ be vertices of an undirected graph $G = (V, E)$, and $\{\{u, v\}, \{u, z\}\} \in E$. Now, suppose execution of Algorithm 3.2. If there exists an orientation $\overrightarrow{uv} \in F$, then it is not possible to make an orientation $\overrightarrow{zu}$, unless $(v, z) \in E$, even if G is not a comparability graph.*

Proof Suppose that $\{v, z\} \notin E$ and $\{u, z\}$ has not been oriented yet. Then addition of $\overrightarrow{uz}$ to set F directly forces execution of the first *for each* loop of the algorithm, where orientation $\overrightarrow{uz}$ is added to the queue and later to F. Existence of the reverse orientation $\overrightarrow{zu} \in F$ immediately stops execution of the algorithm. □

Lemma 3.2 *The number of recursive calls of Algorithm 3.3 for a single vertex is at most equal to its degree.*

Algorithm 3.2 Transitive orientation and simultaneous coloring of a graph

Input: An undirected graph $G = (V, E)$
Output: If G is transitively orientable: heights (coloring) of all vertices

$T \leftarrow V$
$\mathcal{E} \leftarrow E$
$F \leftarrow \emptyset$
Q.clear()
for each $v \in V$ **do** $h[v]$=0
end for
do
  select arbitrary edge $\{u, v\} \in \mathcal{E}$
  Q.push ($\overrightarrow{uv}$)
  while Q is not empty **do**
    $\overrightarrow{uv} = Q$.pop()
    if $\overrightarrow{vu} \in F$ **then**
      Output 'Not a comparability graph'
      return h
    end if
    $F = F \cup \overrightarrow{uv}$
    $\mathcal{E} = \mathcal{E} \setminus \{u, v\}$
    $T[v]$.erase()
    $T[u+1]$.insert(v)
    if $h[u] \leq h[v]$ **then** h = INCHEIGHT(z,u,h,V,T,F)
    for each z=$Adj(u)$ such that $[z \neq Adj(v)$ or $\overrightarrow{vz} \in F]$ **do**
      if $\overrightarrow{uz} \notin \{F \cup Q\}$ **then** Q.push ($\overrightarrow{uz}$)
    end for
    for each z=$Adj(v)$ such that $[z \neq Adj(u)$ or $\overrightarrow{zu} \in F]$ **do**
      if $\overrightarrow{zv} \notin \{F \cup Q\}$ **then** Q.push ($\overrightarrow{zv}$)
    end for
  end while
while $\mathcal{E} \neq \emptyset$
return h

Algorithm 3.3 Increasing of a vertex height during transitive orientation

Input: Vertices u, v, height h, sets V, T, transitive orientation F
Output: Updated heights of u and each its neighbor z such that $\overrightarrow{zu} \in F$

function INCHEIGHT(u,v,h,V,T,F)
  $h[u]$=$h[v]$+1
  for each $z \in T$ such that $z = Adj(u)$ and $\overrightarrow{zu} \in F$ **do**
    if $h[z] \leq h[u]$ **then** h = INCHEIGHT(z,u,h,V,T,F)
  end for
  return h
end function

Proof Let $\{u, v, z\}$ be vertices of an undirected graph $G = (V, E)$, and let $\overrightarrow{zu} \in F$. Let us suppose exploration of an edge $\{u, v\}$ with an orientation $\overrightarrow{uv}$, and let us assume call of function *IncHeight* to increment the height of u. Since $\overrightarrow{zu} \in F$, recursive call of *IncHeight* to increment the height of z is executed. However, from Lemma 3.1 we know, that existence of orientations $\overrightarrow{uv}$ and $\overrightarrow{zu}$ directly forces existence of edge $\{v, z\}$, which means, that v must be a neighbor of z. Furthermore, all recursive calls of *IncHeight* are executed only for neighbors of the initial vertex. Reverse topological ordering T of vertices assures, that particular vertices are updated only once (vertices are visited subsequently, starting from one of the highest ordering which prevents its further exploration). Therefore, the total number of recursive calls of *IncHeight* for the single vertex u of Algorithm 3.2 is at most equal to its degree. □

Finally, we can estimate the computational complexity of the whole algorithm. Clearly, the execution of function *IncHeight* can be done in $O(\Delta(G))$ time. Since each of the *for each* loops is also bounded by $O(\Delta(G))$, we can form the following theorem:

Theorem 3.3 *Comparability graph recognition and simultaneous coloring is bounded by $O(|E| * \Delta(G))$, where $\Delta(G)$ is the maximum degree of a vertex.*

Proof Follows directly from Theorem 3.2, and Lemma 3.2. □

References

1. Aho AV, Hopcroft JE, Ullman J (1983) Data structures and algorithms, 1st edn. Addison-Wesley Longman Publishing Co. Inc, Boston
2. Bang-Jensen J, Gutin GZ (2007) Digraphs: theory, algorithms and applications, 1st edn. Springer, Incorporated
3. Berge C (1973) Graphs and hypergraphs. North-Holland Pub. Co.; American Elsevier Pub. Co., Amsterdam, New York
4. Bron C, Kerbosch J (1973) Algorithm 457: finding all cliques of an undirected graph. Commun ACM 16(9):575–577
5. Chandler D, Chang M-S, Kloks T, Liu J, Peng S-L (2006) Partitioned probe comparability graphs. In: Fomin F (ed) Graph-theoretic concepts in computer science, vol 4271., Lecture notes in computer scienceSpringer, Heidelberg, pp 179–190
6. Corno F, Prinetto P, Sonza Reorda M (1995) Using symbolic techniques to find the maximum clique in very large sparse graphs. In: EDTC '95: Proceedings of the 1995 European conference on design and test, Washington, DC, USA, 1995. IEEE Computer Society, p 320
7. Cornuéjols G, Liu X, Vuskovic K (2003) A polynomial algorithm for recognizing perfect graphs. In: FOCS. IEEE Computer Society Press, pp 20–27
8. Diestel R (2000) Graph theory. Springer, New York (Electronic Edition)
9. Golumbic MC (1977) The complexity of comparability graph recognition and coloring. Computing 18(3):199–208
10. Golumbic MC (1980) Algorithmic graph theory and perfect graphs
11. Grötschel M, Lovász L, Schrijver A (1984) Polynomial algorithms for perfect graphs. Perfect graphs, volume 21 of Annals of Discrete Mathematics. North-Holland, Amsterdam, pp 325–356

12. Habib M, McConnell R, Paul C, Viennot L (2000) Lex-bfs and partition refinement, with applications to transitive orientation, interval graph recognition and consecutive ones testing. Theoret Comput Sci 234(1–2):59–84
13. Harary F (1994) Graph theory. Addison-Wesley, Reading
14. Homer S, Selman A (2001) Computability and complexity theory. In: Texts in computer science. Springer
15. Jensen TR, Toft B (1995) Graph coloring problems. Wiley-Interscience, New York
16. Johnson DS, Papadimitriou CH, Yannakakis M (1988) On generating all maximal independent sets. Inf Process Lett 27(3):119–123
17. Kelly D (1985) Comparability graphs. In: Rival I (ed) Graphs and order, vol 147., NATO ASI seriesSpringer, Netherlands, pp 3–40
18. Kubale M (2004) Graph colorings. American Mathematical Society
19. Lovász L (1972) Normal hypergraphs and the weak perfect graph conjecture. Discrete Math 2:253–267
20. Lovász L (1983) Perfect graphs. In: Beineke L, Wilson R (eds) Selected topics in graph theory, vol 2. Academic Press
21. Makino K, Uno T (2004) New algorithms for enumerating all maximal cliques. In: Hagerup T, Katajainen J (eds) Algorithm theory—SWAT 2004, vol 3111., Lecture notes in computer scienceSpringer, Heidelberg, pp 260–272
22. Shamir R (1994) Advanced topics in graph algorithms
23. Tomita E, Tanaka A, Takahashi H (2006) The worst-case time complexity for generating all maximal cliques and computational experiments. Theoret Comput Sci 363(1):28–42
24. West DB (2000) Introduction to graph theory, 2nd edn. Prentice Hall

Chapter 4
Hypergraphs and Exact Transversals

4.1 Main Definitions and Properties of Hypergraphs

A hypergraph can be seen as a generalization of a graph [1, 4, 7, 8]. Its edges, called hyperedges, may be incident to an arbitrary number of vertices [5, 15].

Definition 4.1 (*Hypergraph*) Hypergraph $\mathcal{H}$ is defined by a pair

$$\mathcal{H} = (V, \mathcal{E}), \tag{4.1}$$

where
$V = \{v_1, \ldots, v_n\}$ is an arbitrary, non-empty set of vertices;
$\mathcal{E} = \{E_1, \ldots, E_m\}$ is a set of hyperedges, subsets of V.

For example, let $V = \{v_1, v_2, v_3, v_4, v_5, v_6\}$ and $\mathcal{E} = \{E_1, E_2, E_3\}$, where $E_1 = \{v_1, v_2, v_3\}$, $E_2 = \{v_1, v_5, v_6\}$ and $E_3 = \{v_4, v_5\}$. Then $\mathcal{H}_1 = \{\{v_1, v_2, v_3\}, \{v_1, v_5, v_6\}, \{v_4, v_5\}\}$ is a hypergraph on V, with the set of hyperedges $\mathcal{E}$.

Similarly to graphs, one of the most popular representations of hypergraphs is *an incidence matrix*, with rows referring to hyperedges, and columns to vertices. If a matrix element equals 1, j-th hyperedge ($j \in 1, \ldots, m$) is incident to i-th vertex ($i \in 1, \ldots, n$). Otherwise the element equals 0

$$A = \begin{cases} 1 & \text{if} \quad v_i \in E_j \\ 0 & \text{if} \quad v_i \notin E_j \end{cases}, \tag{4.2}$$

Incidence matrix $A_{\mathcal{H}_1}$ for hypergraph $\mathcal{H}_1$ is shown in Fig. 4.1. Six rows of the matrix refer to hypergraph vertices and three columns represent its edges.

Definition 4.2 (*Simple hypergraph*) Hypergraph $\mathcal{H} = (V, \mathcal{E})$ is simple if for an arbitrary hyperedge $E_i \in \mathcal{E}$, there is no hyperedge $E_j \in \mathcal{E}$ such that $E_i \subset E_j$.

© Springer International Publishing Switzerland 2017

R. Wiśniewski, *Prototyping of Concurrent Control Systems Implemented in FPGA Devices*, Advances in Industrial Control,
DOI 10.1007/978-3-319-45811-3_4

Fig. 4.1 Incidence matrix $A_{\mathcal{H}_1}$ of hypergraph $\mathcal{H}_1$

$$A_{\mathcal{H}_1} = \begin{array}{c} \begin{array}{cccccc} v_1 & v_2 & v_3 & v_4 & v_5 & v_6 \end{array} \\ \begin{bmatrix} 1 & 1 & 1 & 0 & 0 & 0 \\ 1 & 0 & 0 & 0 & 1 & 1 \\ 0 & 0 & 0 & 1 & 1 & 0 \end{bmatrix} \end{array} \begin{array}{c} E_1 \\ E_2 \\ E_3 \end{array}$$

Let graph $\mathcal{G}(\mathcal{H}) = (V, \mathcal{S})$ be a prime graph of a hypergraph $\mathcal{H} = (V, \mathcal{E})$ with the same set of vertices pairwise connected by edges in such a way that $\mathcal{S} = \{(v, v') \in V^2 : v \neq v' \textit{ and } v, v' \in E \textit{ for some } E \in \mathcal{E}\}$ (cf. [12]).

Definition 4.3 (*Conformal hypergraph*) A hypergraph $\mathcal{H}$ is conformal if all the maximal cliques of its prime graph $\mathcal{G}(\mathcal{H})$ are edges of $\mathcal{H}$ (cf. [5]).

Hypergraph $\mathcal{H}_1$ is simple and conformal. Hypergraph $\mathcal{H}_2 = \{\{v_1, v_2, v_3\}, \{v_1, v_5, v_6\}, \{v_3, v_4\}, \{v_3, v_4, v_5\}\}$ is not simple. Hypergraph $\mathcal{H}_3 = \{\{v_1, v_2, v_3\}, \{v_1, v_5, v_6\}, \{v_3, v_4, v_5\}\}$ is simple but not conformal, since maximal clique $\{v_1, v_3, v_5\}$ of its prime graph is not an edge of $\mathcal{H}_3$.

Definition 4.4 (*Dual hypergraph*) A dual hypergraph $\mathcal{H}^* = (E, \mathcal{V})$ of a hypergraph $\mathcal{H} = (V, \mathcal{E})$ is a hypergraph, whose vertices E correspond to the edges of $\mathcal{H}$, and hyperedges $\mathcal{V}$ correspond to the vertices of $\mathcal{H}$. An incidence matrix $A_{\mathcal{H}^*}$ for dual hypergraph $\mathcal{H}^*$ is the transposed incidence matrix $A_{\mathcal{H}}$ of hypergraph $\mathcal{H}$.

Hypergraph $\mathcal{H}_1^* = (E, \mathcal{V})$ dual to $\mathcal{H}_1$ contains three vertices $E = \{e_1, e_2, e_3\}$ and six hyperedges $\mathcal{V} = \{V_1, \ldots, V_6\}$, where $V_1 = \{e_1, e_2\}$, $V_2 = \{e_2\}$, $V_3 = \{e_1\}$, $V_4 = \{e_3\}$, $V_5 = \{e_2, e_3\}$, and $V_6 = \{e_2\}$. Clearly, $\mathcal{H}_1^*$ is neither simple nor conformal.

Definition 4.5 (*Compatibility relation of vertices*) Two or more vertices of a hypergraph $\mathcal{H}$ are compatible (or in compatibility relation) if they are not connected with any edge.

Definition 4.6 (*Independent set*) A set $I \subseteq V$ is an independent set of a hypergraph $\mathcal{H}$ if all vertices that belong to I are compatible. Set I is called maximal if no $I' \supset I$ is an independent set of $\mathcal{H}$ [9].

Let $\mathcal{I}$ denote the family of all maximal independent sets of the hypergraph $\mathcal{H}$.

Definition 4.7 (*Compatibility hypergraph*) A compatibility hypergraph $\mathcal{H}^C = (V, \mathcal{I})$ of $\mathcal{H}$ is a simple hypergraph with the same set of vertices V, and edge family $\{I\colon I$ is a maximal independent set of $\mathcal{H}\}$.

A compatibility hypergraph of $\mathcal{H}_1$ is a hypergraph $\mathcal{H}_1^C = \{\{v_1, v_4\}, \{v_2, v_4, v_6\}, \{v_3, v_4, v_6\}, \{v_2, v_5\}, \{v_3, v_5\}\}$. Its incidence matrix is shown in Fig. 4.2.

Definition 4.8 (*Transversal*) A transversal [2, 3, 5, 11, 13] of a hypergraph $\mathcal{H} = (V, \mathcal{E})$ is a set $T \subseteq V$ that has nonempty intersection with every edge: $\forall E \in \mathcal{E} : |E \cap T| \geq 1$. *A minimal transversal* is such a transversal which contains no other transversal of $\mathcal{H}$

Fig. 4.2 Incidence matrix of hypergraph $\mathcal{H}_1^C$

$$A_{\mathcal{H}_1^C} = \begin{array}{c} \begin{matrix} v_1 & v_2 & v_3 & v_4 & v_5 & v_6 \end{matrix} \\ \begin{bmatrix} 1 & 0 & 0 & 1 & 0 & 0 \\ 0 & 1 & 0 & 1 & 0 & 1 \\ 0 & 0 & 1 & 1 & 0 & 1 \\ 0 & 1 & 0 & 0 & 1 & 0 \\ 0 & 0 & 1 & 0 & 1 & 0 \end{bmatrix} \begin{matrix} E_1 \\ E_2 \\ E_3 \\ E_4 \\ E_5 \end{matrix} \end{array}$$

A transversal of a hypergraph is also called *hitting set* or *vertex covering* [5].

Definition 4.9 (*Exact transversal*) An exact transversal [9, 17, 18] of a hypergraph $\mathcal{H} = (V, \mathcal{E})$ is a set $X \subseteq V$ that has exactly one intersection with every edge of a hypergraph: $\forall E \in \mathcal{E} : |E \cap X| = 1$.

There are five exact transversals of a hypergraph $\mathcal{H}_1$: $\{\{v_1, v_4\}, \{v_2, v_4, v_6\}, \{v_3, v_4, v_6\}, \{v_2, v_5\}, \{v_3, v_5\}\}$. Please note, that those five sets exactly match edges of compatibility hypergraph of $\mathcal{H}_1$. Furthermore, there are three exact transversals of a hypergraph $\mathcal{H}_1^C$: $\{\{v_1, v_2, v_3\}, \{v_1, v_5, v_6\}, \{v_4, v_5\}\}$, which exactly refer to edges of its primary hypergraph $\mathcal{H}_1$. We will analyze this interesting property later.

Let us now point out some properties of hypergraphs that are obvious and well known (cf. [5, 6, 9, 12, 13, 17]):

Corollary 4.1 *Let $\mathcal{I}$ denote the family of all maximal independent sets of the hypergraph $\mathcal{H} = (V, \mathcal{E})$. Then each independent set has at most one intersection with every edge $E \in \mathcal{E}$ of a hypergraph: $\forall\, I \in \mathcal{I} : |I \cap E| \leq 1$.*

Corollary 4.2 *Each exact transversal of $\mathcal{H}$ is also an independent set of $\mathcal{H}$.*

Corollary 4.3 *Let $\mathcal{X}$ denote the family of all exact transversals of the hypergraph $\mathcal{H}$, and let $\mathcal{H}^C = (V, \mathcal{I})$ denote the compatibility hypergraph of $\mathcal{H}$. Then $\mathcal{X} \subseteq \mathcal{I}$.*

In 1994 an essential paper regarding exact transversals was published. In [9] it was proved that subsequent minimal exact transversals in an exact transversal hypergraph are computed in polynomial time. However, the number of all minimal exact transversals may be exponential, thus calculation of all solutions is also classified as exponential.

Definition 4.10 (*Exact transversal hypergraph, xt-hypergraph*) An *exact transversal hypergraph* (or just *xt-hypergraph*) is a hypergraph in which all minimal transversals are also exact transversals [9].

An extension of an *exact transversal hypergraph* is another structure, called *c-exact hypergraph*, or just *exact hypergraph*. It was initially used in [19] and formally defined in [18].

Definition 4.11 (*Compatible exact hypergraph, c-exact hypergraph*) A *compatible exact hypergraph* (or just *c-exact hypergraph*) is a hypergraph in which any set of compatible vertices (not connected with an edge) belongs to at least one exact transversal.

Hypergraph $\mathcal{H}_1$ is a compatible exact hypergraph. Moreover, its compatibility hypergraph, $\mathcal{H^C}_1$ is also c-exact.

A c-exact hypergraph is a generalization of an xt-hypergraph. In this case, the condition that each minimal transversal is also an exact transversal does not have to be satisfied. However, any subset of vertices which are in a compatibility relation (i.e., not connected by any edge) must belong to at least one exact transversal. Hence, each vertex of the c-exact hypergraph must belong to at least one exact transversal.

Clearly, both types of hypergraphs (*xt-hypergraph* and *c-exact hypergraph*) are simple. Notice, that empty sets such as $\emptyset$ and $\{\emptyset\}$ are also classified as *xt* and *c-exact* hypergraphs.

Next sections present main properties of c-exact hypergraphs in more details. Furthermore, computational complexity of some problems related to the c-exact hypergraphs is analyzed and discussed.

4.2 Properties of C-Exact Hypergraphs

In this section the most interesting properties of c-exact hypergraphs, that have been never published in the literature before are presented.

At the beginning, we show that the compatibility hypergraph of a c-exact hypergraph can be achieved as a calculation of all exact transversals. Moreover, if the c-exact hypergraph is conformal, its compatibility hypergraph is c-exact, as well.

Theorem 4.1 *Let $\mathcal{H} = (V, \mathcal{E})$ be a c-exact hypergraph. Then each maximal independent set I of $\mathcal{H}$ is also an exact transversal of $\mathcal{H}$.*

Proof From the definition of c-exact hypergraph we know that each set of compatible vertices form at least one exact transversal. This means that each independent set I of $\mathcal{H}$ must belong to at least one exact transversal X of $\mathcal{H}$, in such a way, that $I \subseteq X$. Thus, I is maximal if and only if $I = X$. □

Theorem 4.2 *Let $\mathcal{H} = (V, \mathcal{E})$ be a c-exact hypergraph, let $\mathcal{H^C}$ denote the compatibility hypergraph of $\mathcal{H}$, and let $\mathcal{X}$ denote the family of all exact transversals of $\mathcal{H}$. Then $\mathcal{H^C} = (V, \mathcal{X})$.*

Proof Follows directly from Corollaries 4.2, 4.3, and Theorem 4.1.

From the above theorems and Corollary 4.1, we can easily observe the following properties of a c-exact hypergraph:

Corollary 4.4 *Let $\mathcal{I}$ denote the family of all maximal independent sets of the c-exact hypergraph $\mathcal{H} = (V, \mathcal{E})$. Then each independent set has exactly one intersection with every edge $E \in \mathcal{E}$ of a hypergraph: $\forall\, I \in \mathcal{I} : |I \cap E| = 1$.*

Since each maximal independent set of a c-exact hypergraph is also an exact transversal, we immediately have

Corollary 4.5 *Let $\mathcal{X}$ denote the family of all exact transversals of the c-exact hypergraph $\mathcal{H} = (V, \mathcal{E})$. Then each exact transversal has exactly one intersection with every edge $E \in \mathcal{E}$ of a hypergraph: $\forall X \in \mathcal{X} : |X \cap E| = 1$.*

Theorem 4.3 *Let $\mathcal{H} = (V, \mathcal{E})$ be a c-exact and conformal hypergraph, let $\mathcal{H}^C = (V, \mathcal{X})$ denote the compatibility hypergraph of $\mathcal{H}$, and let $(\mathcal{H}^C)^C$ denote the compatibility hypergraph of $\mathcal{H}^C$. Then $\mathcal{H} = (\mathcal{H}^C)^C$.*

Proof From the definition of an exact transversal, we know that X is exact if $\forall E \in \mathcal{E} : |E \cap X| = 1$. On the other hand, from Corollary 4.5 we have that $\forall X \in \mathcal{X} : |X \cap E| = 1$. It means that each set $E \in \mathcal{E}$ is an exact transversal (and maximal independent set) of $\mathcal{H}^C$. Since $\mathcal{H}$ is conformal, no more independent sets exist in $\mathcal{H}^C$ and $\mathcal{H} = (\mathcal{H}^C)^C$.□

Theorem 4.4 *Let $\mathcal{H} = (V, \mathcal{E})$ be a c-exact and conformal hypergraph, let $\mathcal{H}^C$ denote the compatibility hypergraph of $\mathcal{H}$. Then $\mathcal{H}^C = (V, \mathcal{X})$ is also c-exact.*

Proof In the proof of the Theorem 4.3 we have shown that each hyperedge of $\mathcal{H}$ refers to an exact transversal (and maximal independent set) of $\mathcal{H}^C$. Clearly, since $\mathcal{H}$ is conformal, no more independent sets exist in $\mathcal{H}^C$. Therefore, any set of compatible vertices in the hypergraph $\mathcal{H}^C$ must form an exact transversal, which means that $\mathcal{H}^C$ is c-exact. □

Hypergraph $\mathcal{H}_1$ is c-exact and conformal. Therefore, its compatibility hypergraph $\mathcal{H}_1^C$ is also c-exact. Furthermore, sice $\mathcal{H}_1^C$ is conformal, its compatibility hypergraph $(\mathcal{H}_1^C)^C$ should be also c-exact. In a fact, a hypergraph $(\mathcal{H}_1^C)^C = \mathcal{H}_1$ is c-exact.

4.3 Algorithms Related to C-Exact Hypergraphs

This section focusses on the computational complexity of algorithms related to c-exact hypergraphs. Such properties were initially described in [18]. Now, we shall extend them and prove formally.

At the beginning we will show that subsequent exact transversals of a c-exact hypergraph can be computed in polynomial time. Notice, that due to the unique properties of c-exact hypergraphs, such a problem can be presented in a different way (like maximal independent set/minimal transversal computational complexity, cf. [6, 10, 16, 18]).

Let us introduce additional notation. A reduced (or induced) hypergraph (by a vertex v_r) $\mathcal{H}_{-v_r} = (V', \mathcal{E}')$ of $\mathcal{H} = (V, \mathcal{E})$ is a hypergraph with reduced sets of hyperedges and vertices: $\forall E \in \mathcal{E}\ (E \not\ni v_r \Rightarrow E \in \mathcal{E}')$ and $\forall v \in V\ (\nexists E \in \mathcal{E} \setminus \mathcal{E}' : v \in E \Rightarrow v \in V')$. In other words, $\mathcal{H}_{-v_r}$ contains all edges of initial hypergraph $\mathcal{H}$ that do not contain vertex v_i. Furthermore, $\mathcal{H}_{-v_r}$ contains only these vertices of $\mathcal{H}$ that do not belong to any of reduced edges. Hypergraph $\mathcal{H}_{-v_r}$ can be simply achieved by removing all edges that contain vertex v_i and all vertices that are incident to them.

Lemma 4.1 *Let $\mathcal{H} = (V, \mathcal{E})$ be a c-exact hypergraph, then the reduced hypergraph (by a vertex v_r) $\mathcal{H}_{-v_r} = (V', \mathcal{E}')$ is also c-exact.*

Proof If $\mathcal{H}_{-v_r} = \{\emptyset\}$, the hypergraph is obviously c-exact, so for the rest, we exclude this case.

It is clear, that reduced hypergraph $\mathcal{H}_{-v_r}$ contains only those vertices of $\mathcal{H}$ that are compatible with v_r. All hyperedges that contain v_r are removed and all vertices that belong to those edges are removed as well.

On the other hand, all remaining vertices of $\mathcal{H}_{-v_r}$ could not belong to any of the removed edges. Obviously, they are adjacent to the same hyperedges as before reduction. Moreover, those edges also remain untouched. Furthermore, it is easy to see, that compatibility relations between any set of vertices in the hypergraph $\mathcal{H}_{-v_r}$ remain unchanged. Clearly, any transversal $X' = X \backslash \{v_r\}$ of $\mathcal{H}_{-v_r}$ formed during the reduction is still exact. Since $\mathcal{H}$ is a c-exact hypergraph, where any set of compatible vertices belong to at least one exact transversal, the reduced hypergraph $\mathcal{H}_{-v_r}$ must also be c-exact. □

Lemma 4.2 *An exact transversal in a c-exact hypergraph can be computed in polynomial time.*

Proof Consider a method shown in Algorithm 4.1 that generates a single exact transversal X_{v_i} in a hypergraph $\mathcal{H}$ containing n vertices and m edges. The algorithm starts from a vertex v_i.

Algorithm 4.1 Computation of a single exact transversal

Input: A c-exact hypergraph $\mathcal{H}$
Output: An exact transversal X_{v_i}

```
X_{v_i} = ∅
v = v_i
while H ≠ ∅ do
    X_{v_i} = X_{v_i} ∪ {v}
    H = H_{-v_i}
    if H ≠ ∅ then
        select any vertex v ∈ V
    end if
end while
return X_{v_i}
```

From Lemma 4.1 it is easy to see that Algorithm 4.1 permits us to find a single transversal in a c-exact hypergraph. The consecutive vertices are added to the transversal X_{v_i} at each execution of the *while* loop. This process is repeated until the hypergraph is completely reduced, which means that set X_{v_i} forms an exact cover (cf. [14]).

The main loop of the presented algorithm is executed at most n times. Clearly, the operation $\mathcal{H} = \mathcal{H}_{-v_r}$ (reduction of hyperedges and all vertices that belong to them) is computable in time $O(mn)$. Thus, the runtime of the whole algorithm is bounded by $O(mn^2)$. □

Theorem 4.5 *Subsequent exact transversals in a c-exact hypergraph can be computed in polynomial time.*

To clarify the proof, we will operate on the algorithm that generates all subsequent exact transversals. Notice that Algorithm 4.2 calculates all exact transversals in a c-exact hypergraph. The subsequent transversals are generated in polynomial time. Notice, that the total number of all exact transversals in a hypergraph can be exponential [9].

Proof Let us extend previous method of a single exact transversal computation. Algorithm 4.2 generates subsequent exact transversals in a hypergraph $\mathcal{H}$ containing n vertices and m edges.

Algorithm 4.2 Computation of subsequent exact transversals

Input: A c-exact hypergraph $\mathcal{H}$, partial exact transversal X_d
Output: Subsequent exact transversals

if $\mathcal{H} = \emptyset$ **then**
    output X_d
    return
else
    select edge E_e that contains the fewest vertices
    for all v_i such that $v_i \in E_e$ **do**
        $X_d = X_d \cup \{v_i\}$
        $\mathcal{H} = \mathcal{H}_{-v_i}$
        call recursively ***Algorithm 3.2***
end if

The presented method operates in the similar manner as Algorithm 4.1, however after finding the first exact transversal, it still looks for the subsequent transversals. The selection of an edge with the fewest vertices and loop *for* are essential steps of the algorithm. The subsequent recurrences are executed for the subsequent vertices v_i that belong to E_e. Notice, that any pair of vertices $v_i \in E_e$ are incompatible, because they are connected by an edge. Simply, the selection process of an edge that contains the fewest vertices can be executed in time $O(mn)$ (we assume that calculation of the number of vertices in each edge is needed). As it was already shown (see proof of Lemma 4.2), the reduction $\mathcal{H} = \mathcal{H}_{-v_i}$ is also performed in time $O(mn)$. Thus, the runtime of a single recurrence is quadratic ($O(mn)$).

The maximum number of recursive calls is equal to n. If hypergraph $\mathcal{H}$ is empty, it means that the transversal was found and the algorithm recursively returns to a higher level. Summarizing, it is easy to see, that the next exact transversal will be computed in a linear recursive calls of the algorithm ($O(n)$). Thus, the subsequent transversals are calculated in time $O(mn^2)$. This means that the subsequent exact transversals in a c-exact hypergraph can be computed in polynomial time. □

The method is similar to the general algorithm of exact covering calculation, proposed by Knuth in [14], where effective implementation in programming languages is presented as well. Knuth proved that such an algorithm enables finding all the exact transversals in any hypergraph (cf. [14]). We showed, that in the case of exact hypergraphs, the subsequent exact transversals are found in polynomial time, which is not possible in general case of hypergraphs [5, 9].

References

1. Anderson M, Meyer B, Olivier P (2011) Diagrammatic representation and reasoning. Springer Science & Business Media
2. Bailey J, Manoukian T, Ramamohanarao K (2003) A fast algorithm for computing hypergraph transversals and its application in mining emerging patterns. In: ICDM, IEEE Computer Society, pp 485–488
3. Berge C (1973) Graphs and hypergraphs. North-Holland Pub. Co., American Elsevier Pub. Co., Amsterdam, New York
4. Berge C (1984) Hypergraphs: combinatorics of finite sets, vol 45. Elsevier
5. Berge C (1989) Hypergraphs: combinatorics of finite sets, North-Holland
6. Boros E, Gurvich V, Khachiyan L, Makino K (2000) An efficient incremental algorithm for generating all maximal independent sets in hypergraphs of bounded dimension. Parallel Process Lett 10:253–266
7. Bothorel C, Bouklit M (2011) An algorithm for detecting communities in folksonomy hypergraphs. In: I2CS 2011: 11th international conference on innovative internet community services, LNI, pp 159–168
8. Dekker L, Smit W, Zuidervaart J (2013) In: Massively parallel processing applications and development: proceedings of the 1994 EUROSIM conference on massively parallel processing applications and development, Delft, The Netherlands, 21–23 June 1994, Elsevier
9. Eiter T (1994) Exact transversal hypergraphs and application to boolean μ-functions. J Symb Comput 17(3):215–225
10. Eiter T, Gottlob G (1995) Identifying the minimal transversals of a hypergraph and related problems. SIAM J Comput 24(6):1278–1304
11. Eiter T, Gottlob G, Makino K (2002) New results on monotone dualization and generating hypergraph transversals. SIAM J Comput 14–22
12. Hodkinson I, Otto M (2003) Finite conformal hypergraph covers and gaifman cliques in finite structures. Bull Symb Logic 9:387–405
13. Kavvadias D, Stavropoulos EC (1999) Evaluation of an algorithm for the transversal hypergraph problem. In Proceedings of the 3rd international workshop on algorithm engineering, Springer, London, UK, pp 72–84
14. Knuth D (2000) Dancing links. In: Millennial perspectives in computer science, Palgrave, pp 187–214
15. Lovász L (1972) Normal hypergraphs and the weak perfect graph conjecture. Discret Math 2:253–267
16. Mishra N, Pitt L (1997) Generating all maximal independent sets of bounded-degree hypergraphs. In: COLT '97: Proceedings of the 10th annual conference on Computational learning theory, ACM, New York, NY, USA, pp 211–217
17. Rauf I (2011) Polynomially solvable cases of hypergraph transversal and related problems. Doctoral dissertation, Universität des Saarlandes, Saarbrücken, Oct 2011

18. Wiśniewska M (2012) Application of hypergraphs in decomposition of discrete systems, vol 23. Lecture Notes in Control and Computer Science University of Zielona Góra Press, Zielona Góra
19. Wiśniewska M, Adamski M, Wiśniewski R (2011) Exact transversals in decomposition of Petri nets into concurrent subnets. Meas Autom Monit 58(8):851–853 in Polish

Chapter 5
Analysis of Concurrent Control Systems

5.1 State Equation and Place Invariants

The dynamic behavior of Petri nets can be expressed by the set of algebraic equations [24]. In such an approach, the net is represented by a matrix. Particular markings (represented by nonnegative integers) can be achieved by solving the linear equations [19, 21, 28]. At the beginning, let us introduce necessary definitions [9, 15, 19–21, 24, 28]. Please note, that all presented notations refer to pure and safe Petri nets.

Definition 5.1 (*Incidence matrix of a Petri net*) An *incidence matrix* of a Petri net $PN = (P, T, F, M_0)$ with $n = |P|$ places and $m = |T|$ transitions is an $A_{m \times n}$ matrix (where m refers to rows, and n refers to columns) of integers, given by

$$a_{ij} = \begin{cases} -1, & (p_i, t_j) \in F \\ 1, & (t_j, p_i) \in F \\ 0, & \text{otherwise} \end{cases} .$$

A cell a_{ij} of matrix A is connected with place p_i and transition t_j. The columns of the matrix correspond to places, while the rows refer to transitions of a Petri net (in some notations such values are reversed).

Recall Petri net PN_1, presented in Fig. 2.1. The incidence matrix A_{PN_1} for this net is shown in Fig. 5.1.

It is easy to notice that i-th row of the incidence matrix indicate the input and output places for the transition i. For example, transition t_2 moves tokens from places p_2 and p_4 to p_1 and p_5. Similarly, j-th column of the matrix refers to the input and output transitions for the place p_j. For example, place p_3 achieves a token by firing transition t_1, while execution of t_3 removes it.

© Springer International Publishing Switzerland 2017

R. Wiśniewski, *Prototyping of Concurrent Control Systems Implemented in FPGA Devices*, Advances in Industrial Control,
DOI 10.1007/978-3-319-45811-3_5

Fig. 5.1 Incidence matrix A_{PN_1} for Petri net PN_1

$$A_{PN_1} = \begin{array}{c} \\ \\ \\ \end{array} \begin{array}{cccccc} p_1 & p_2 & p_3 & p_4 & p_5 & p_6 \end{array}$$

$$A_{PN_1} = \begin{bmatrix} -1 & 0 & 1 & 1 & 0 & -1 \\ 1 & -1 & 0 & -1 & 1 & 0 \\ 0 & 1 & -1 & 0 & -1 & 1 \end{bmatrix} \begin{array}{c} t_1 \\ t_2 \\ t_3 \end{array}$$

From the mathematical point of view, k-th marking (denoted as M_k) in a safe Petri net can be seen as a $n \times 1$ binary vector $\overrightarrow{M_k}$, where n means the number of places of the net. The i-th entry of $\overrightarrow{M_k}$ indicates whenever the place is marked after k-th firing of transitions in firing sequence. We shall denote the particular k-th firing by σ_k (and a corresponding binary vector by $\overrightarrow{\sigma_k}$), while the whole firing sequence is marked by σ. Changes between consecutive markings may be calculated from the linear equation

$$\overrightarrow{M_k} = \overrightarrow{M_{k-1}} + A^T \bullet \overrightarrow{\sigma_k}, \tag{5.1}$$

where M_k is the destination marking (reachable from M_{k-1}), and A^T is transposed incidence matrix of a Petri net.

Let us explain the above equation by an example. Recall all the reachable markings of PN_1, shown in Fig. 2.2. Such markings can be expressed by the binary vectors $\overrightarrow{M_0} = [110001]^T$, $\overrightarrow{M_1} = [011100]^T$, $\overrightarrow{M_2} = [101010]^T$. Assume, that we try to compute the value of state M_2. Such a marking is reachable from M_1 by firing t_2, which means that $\overrightarrow{\sigma_2} = [010]^T$ for $T = \{t_1, t_2, t_3\}$. Thus, the state equation can be expressed as follows:

$$\overrightarrow{M_1} + A^T \bullet \overrightarrow{\sigma_2} = \begin{bmatrix} 0 \\ 1 \\ 1 \\ 1 \\ 0 \\ 0 \end{bmatrix} + \begin{bmatrix} -1 & 1 & 0 \\ 0 & -1 & 1 \\ 1 & 0 & -1 \\ 1 & -1 & 0 \\ 0 & 1 & -1 \\ -1 & 0 & 1 \end{bmatrix} \bullet \begin{bmatrix} 0 \\ 1 \\ 0 \end{bmatrix} = \begin{bmatrix} 0 \\ 1 \\ 1 \\ 1 \\ 0 \\ 0 \end{bmatrix} + \begin{bmatrix} 1 \\ -1 \\ 0 \\ -1 \\ 1 \\ 0 \end{bmatrix} = \begin{bmatrix} 1 \\ 0 \\ 1 \\ 0 \\ 1 \\ 0 \end{bmatrix} = \overrightarrow{M_2}.$$

Finally, we can formally define the state equation of a Petri net for the destination marking M_k that is reachable from M_0.

Definition 5.2 (*State equation of a Petri net*) A *state equation* of a Petri net is defined as a linear equation

$$\overrightarrow{M_k} = \overrightarrow{M_0} + A^T \bullet \overrightarrow{x} \tag{5.2}$$

where:

- M_0 is an initial marking,
- M_k is a destination marking, that is reachable from M_0,
- A^T is a transposed incidence matrix of a Petri net,
- $\overrightarrow{x}$ is the Parikh vector of a firing sequence σ such that $M_0 \sigma M_k$.

For the net PN_1 computation of the state M_2 from the initial marking M_0 requires a firing sequence of transitions t_1 and t_2, thus $\overrightarrow{x} = [110]^T$. The state equation for M_2 is calculated as follows:

$$\overrightarrow{M_0} + A^T \bullet \overrightarrow{x} = \begin{bmatrix} 1 \\ 1 \\ 0 \\ 0 \\ 0 \\ 1 \end{bmatrix} + \begin{bmatrix} -1 & 1 & 0 \\ 0 & -1 & 1 \\ 1 & 0 & -1 \\ 1 & -1 & 0 \\ 0 & 1 & -1 \\ -1 & 0 & 1 \end{bmatrix} \bullet \begin{bmatrix} 1 \\ 1 \\ 0 \end{bmatrix} = \begin{bmatrix} 1 \\ 0 \\ 1 \\ 0 \\ 1 \\ 0 \end{bmatrix} = \overrightarrow{M_2}.$$

Definition 5.3 (*Place invariant*) *Place invariant (p-invariant)* of a Petri net *PN* is a vector $\overrightarrow{y}$ of nonnegative integers that solves the equation

$$\overrightarrow{y} \bullet A^T = 0, \tag{5.3}$$

where $\overrightarrow{y} \neq 0$ and A is an incidence matrix of a Petri net. Each entry of $\overrightarrow{y}$ refers to a place of *PN*. The set of places that correspond to nonzero values of a p-invariant is called its *support* and shall be denoted by I. A p-invariant is *minimal* if no proper nonempty subset of its support is a support of another p-invariant [20, 24].

Place invariants computation is nowadays one of the most popular analysis methods of a Petri net. Among the others, p-invariants are used to validate important properties of Petri nets, such as boundness and necessary condition of liveness [20, 24]. Furthermore, they can be applied to the prototyping flow of concurrent systems in various areas, like manufacturing [11], robot-control applications [22] or transportation [13, 23]. In this book we shall show that p-invariants are also useful in prototyping of concurrent controllers implemented in reprogrammable devices, such as FPGAs.

Let us now present the most popular algorithm of place invariant computation. The method was initially shown in [20]. It is based on the integer linear algebra, and permits to obtain all the p-invariants of a Petri net.
The whole procedure can be divided into the following steps [20]:

1. *Initialization*: Form a unit matrix $Q = [D|A^T]$. Initially, D is equal to an identity matrix I_n, where n is the number of places of the Petri net. Matrix A^T is a transposed incidence matrix of a Petri net with rows corresponding to places $i = \{1, \dots, n\}$ and columns $j = \{1, \dots, m\}$ referring to the transitions of Petri net, where m is

equal to the number of transitions of the Petri net. Matrix Q is a base for further linear computation.

2. *Minimal invariants computation*: For each column j repeat the procedure:
   (a) Find row pairs that annul the j-th column (i.e., their sum is equal to 0) and append it to the matrix Q.
   (b) Delete all rows of Q in which the intersection with j-th column is not equal to 0.
   (c) Eliminate non-minimal invariants by reducing redundant rows of Q (i.e., rows that binary cover the other ones).

The above algorithm assures obtaining all the minimal invariants in a Petri net [20]. The final results can be achieved from the matrix D, which rows correspond to the obtained invariants.

It is assumed, that only minimal place invariants are taken into account in case of analysis and decomposition of concurrent control systems described in this book. Therefore, from this point, short notation *place invariant* refers to a *minimal place invariant*, unless stated otherwise.

Let us explain the algorithm by an example. Figure 5.2 shows the initial unit matrix Q for the Petri net PN_1. There is also current support value of invariants presented.

At the first interaction of the algorithm, transition t_1 is examined. Four rows (first, third, fourth, and sixth) of the matrix are removed from Q, while row pairs that annul the first column (t_1) are appended to Q. Notice, that the second and the fifth rows of the matrix remain untouched. The result of such an operation is illustrated in Fig. 5.3.

Transition t_2 is checked at the second interaction of the algorithm. Similarly to the examination of t_1, four rows (first, second, third, and the last one) are removed from the matrix, and four new values are added to Q. Figure 5.4 shows the resultant matrix. Notice, that the last row binary covers the first and the second one. Therefore, it is reduced from the matrix Q.

At the third interaction, the last transition is examined. No further operation is performed, since intersection of t_3 with all the rows is equal to 0. Finally, five place invariants have been obtained

- $\vec{y_1} = [100100]$, (with support of the set of places) $I_1 = \{p_1, p_4\}$,
- $\vec{y_2} = [001001]$, $I_2 = \{p_3, p_6\}$,

$$Q = \begin{array}{c} \\ \end{array}\begin{array}{cccccc|ccc} p_1 & p_2 & p_3 & p_4 & p_5 & p_6 & t_1 & t_2 & t_3 \\ 1 & 0 & 0 & 0 & 0 & 0 & -1 & 1 & 0 \\ 0 & 1 & 0 & 0 & 0 & 0 & 0 & -1 & 1 \\ 0 & 0 & 1 & 0 & 0 & 0 & 1 & 0 & -1 \\ 0 & 0 & 0 & 1 & 0 & 0 & 1 & -1 & 0 \\ 0 & 0 & 0 & 0 & 1 & 0 & 0 & 1 & -1 \\ 0 & 0 & 0 & 0 & 0 & 1 & -1 & 0 & 1 \end{array} \qquad \begin{array}{c} \text{current support} \\ \{p_1\} \\ \{p_2\} \\ \{p_3\} \\ \{p_4\} \\ \{p_5\} \\ \{p_6\} \end{array}$$

Fig. 5.2 The initial matrix Q of Martinez–Silva algorithm for Petri net PN_1

$$
Q=\begin{array}{c}
\begin{array}{cccccc|ccc} p_1 & p_2 & p_3 & p_4 & p_5 & p_6 & t_1 & t_2 & t_3 \end{array}\\
\left[\begin{array}{cccccc|ccc}
0 & 1 & 0 & 0 & 0 & 0 & 0 & -1 & 1\\
0 & 0 & 0 & 0 & 1 & 0 & 0 & 1 & -1\\
1 & 0 & 1 & 0 & 0 & 0 & 0 & 1 & -1\\
1 & 0 & 0 & 1 & 0 & 0 & 0 & 0 & 0\\
0 & 0 & 1 & 0 & 0 & 1 & 0 & 0 & 0\\
0 & 0 & 0 & 1 & 0 & 1 & 0 & -1 & 1
\end{array}\right]
\end{array}
\qquad
\begin{array}{c}
\text{current support}\\
\{p_2\}\\
\{p_5\}\\
\{p_1,p_3\}\\
\{p_1,p_4\}\\
\{p_3,p_6\}\\
\{p_4,p_6\}
\end{array}
$$

Fig. 5.3 Matrix Q of after the first interaction of an algorithm

$$
Q=\begin{array}{c}
\begin{array}{cccccc|ccc} p_1 & p_2 & p_3 & p_4 & p_5 & p_6 & t_1 & t_2 & t_3 \end{array}\\
\left[\begin{array}{cccccc|ccc}
1 & 0 & 0 & 1 & 0 & 0 & 0 & 0 & 0\\
0 & 0 & 1 & 0 & 0 & 1 & 0 & 0 & 0\\
0 & 1 & 0 & 0 & 1 & 0 & 0 & 0 & 0\\
1 & 1 & 1 & 0 & 0 & 0 & 0 & 0 & 0\\
0 & 0 & 0 & 1 & 1 & 1 & 0 & 0 & 0\\
1 & 0 & 1 & 1 & 0 & 1 & 0 & 0 & 0
\end{array}\right]
\end{array}
\qquad
\begin{array}{c}
\text{support}\\
\{p_1,p_4\}\\
\{p_3,p_6\}\\
\{p_2,p_5\}\\
\{p_1,p_2,p_3\}\\
\{p_4,p_5,p_6\}\\
\{p_1,p_3,p_4,p_6\}
\end{array}
$$

Fig. 5.4 The output matrix Q of Martinez–Silva algorithm for Petri net PN_1

- $\overrightarrow{y_3} = [010010]$, $I_3 = \{p_2, p_5\}$,
- $\overrightarrow{y_4} = [111000]$, $I_4 = \{p_1, p_2, p_3\}$,
- $\overrightarrow{y_5} = [000111]$, $I_5 = \{p_4, p_5, p_6\}$.

Let us now present invariant analysis result for the other concurrent control systems. A model of milling machine presented in Fig. 2.4 and described by the interpreted Petri net PN_2 results in eight invariants

- $\overrightarrow{y_1} = [111001000000000000011]$, $I_1 = \{p_1, p_2, p_3, p_6, p_{20}, p_{21}\}$,
- $\overrightarrow{y_2} = [100111000000000000011]$, $I_2 = \{p_1, p_4, p_5, p_6, p_{20}, p_{21}\}$,
- $\overrightarrow{y_3} = [111000111111100000011]$, $I_3 = \{p_1, p_2, p_3, p_7, \ldots, p_{13}, p_{20}, p_{21}\}$,
- $\overrightarrow{y_4} = [100110111111100000011]$, $I_4 = \{p_1, p_4, p_5, p_7, \ldots, p_{13}, p_{20}, p_{21}\}$,
- $\overrightarrow{y_5} = [111000000000011100011]$, $I_5 = \{p_1, p_2, p_3, p_{14}, p_{15}, p_{16}, p_{20}, p_{21}\}$,
- $\overrightarrow{y_6} = [100110000000011100011]$, $I_6 = \{p_1, p_4, p_5, p_{14}, p_{15}, p_{16}, p_{20}, p_{21}\}$,
- $\overrightarrow{y_7} = [111000000000000011111]$, $I_7 = \{p_1, p_2, p_3, p_{17}, p_{18}, p_{19}, p_{20}, p_{21}\}$,
- $\overrightarrow{y_8} = [100110000000000011111]$, $I_8 = \{p_1, p_4, p_5, p_{17}, p_{18}, p_{19}, p_{20}, p_{21}\}$.

Furthermore, three p-invariants are achieved during analysis of the concurrent controller of traffic light system illustrated by the net PN_3 from Fig. 2.6

- $\overrightarrow{y_1} = [011100]$, $I_1 = \{p_2, p_3, p_4\}$,
- $\overrightarrow{y_2} = [000011]$, $I_2 = \{p_5, p_6\}$,
- $\overrightarrow{y_3} = [111010]$, $I_3 = \{p_1, p_2, p_3, p_5\}$.

Place invariant usually (but not always) indicate sequential areas of the prototyped control systems. Thus, they are closely related to the state machine components. We

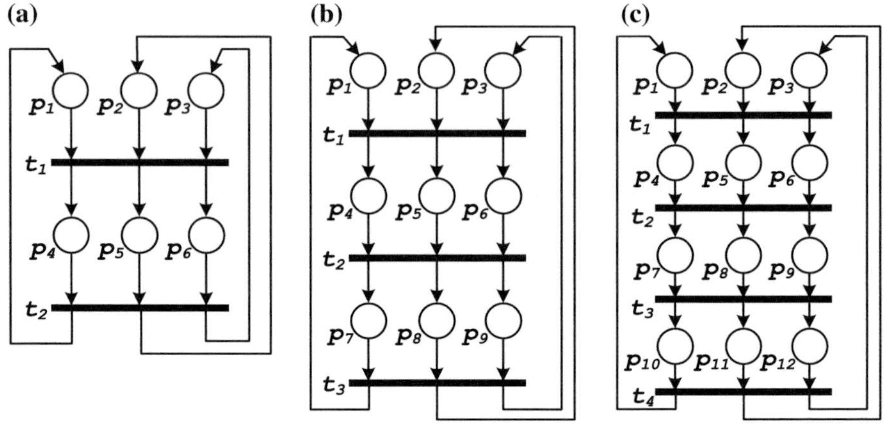

Fig. 5.5 Example of exponential explosion in place invariants computation

shall enhance such a property in Sect. 5.3, where the sequentiality relation between places in a Petri net is analyzed.
The main bottleneck of place invariants analysis is exponential computational complexity [20, 33]. Let us illustrate such a problem by an example. Consider three Petri nets shown in Fig. 5.5. The left-most one (a) results in nine place invariants. Its slight modified version is presented in the middle (b). Addition of three places and one transition extends the number of achieved p-invariants to 27. Further enhancement of the net by additional three places (c) results in 81 place invariants.

5.2 Concurrency Analysis

Recall Definition 2.16 from Chap. 2. It says, that two places of an interpreted Petri net are *concurrent* if there exists a state, where both places are marked simultaneously. Such a property can be easily and efficiently represent by an undirected graph. Let us define such a structure.

Definition 5.4 (*Concurrency graph*) *Concurrency graph* $G_C = (P, E)$ of an interpreted Petri net $PN = (P, T, F, M_0, X, Y)$ is a graph which vertices refer to the places of the net. Two vertices are connected by an edge if corresponding places are concurrent in PN, i.e., there exists a reachable marking, where both places are simultaneously marked. The relation represented by the set of edges is the *concurrency relation* and is denoted as $||$.

Figure 5.6 (left) shows an exemplary *concurrency graph* G_{C_1} for Petri net PN_1 (from Fig. 2.1) and its neighborhood matrix (right). Six vertices correspond to the places of the Petri net. Each two vertices are connected by an edge if they are concurrent. There are totally nine edges in G_{C_1}, that refer to the *concurrency relation*.

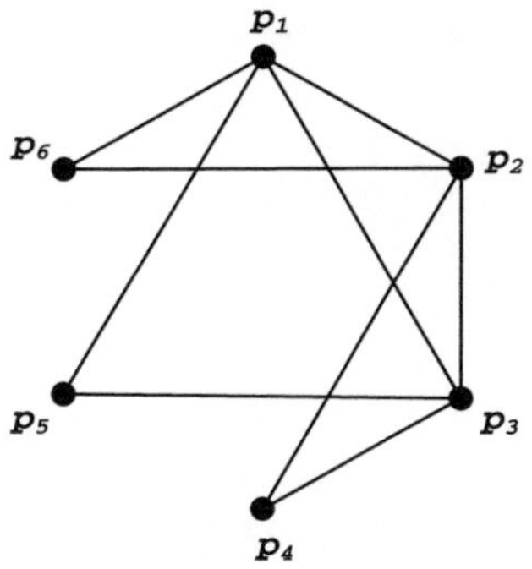

$$N_{G_{C_1}} = \begin{array}{c} \begin{array}{cccccc} p_1 & p_2 & p_3 & p_4 & p_5 & p_6 \end{array} \\ \left[\begin{array}{cccccc} 0 & 1 & 1 & 0 & 1 & 1 \\ 1 & 0 & 1 & 1 & 0 & 1 \\ 1 & 1 & 0 & 1 & 1 & 0 \\ 0 & 1 & 1 & 0 & 0 & 0 \\ 1 & 0 & 1 & 0 & 0 & 0 \\ 1 & 1 & 0 & 0 & 0 & 0 \end{array}\right] \end{array} \begin{array}{l} \\ p_1 \\ p_2 \\ p_3 \\ p_4 \\ p_5 \\ p_6 \end{array}$$

Fig. 5.6 Concurrency graph for PN_1 (*left*) and its neighborhood matrix (*right*)

Such a relation can be clearly illustrated by an adjacency matrix of the concurrency graph, as it is presented in Fig. 5.6 (right). For example place p_1 is concurrent to four places: p_2, p_3, p_5, p_6, which can be shortly denoted as $p_1||p_2, p_1||p_3, p_1||p_5, p_1||p_6$.

More complicated example is shown in Fig. 5.7, where adjacency matrix of concurrency graph G_{C_2} for the net PN_2 is presented. There are 20 places in the net that correspond to the rows and columns of the matrix. Please note that concurrency graph consists of $|E| = 68$ edges.

$$N_{G_{C_2}} = \begin{array}{c} \begin{array}{ccccccccccccccccccccc} p_1 & p_2 & p_3 & p_4 & p_5 & p_6 & p_7 & p_8 & p_9 & p_{10} & p_{11} & p_{12} & p_{13} & p_{14} & p_{15} & p_{16} & p_{17} & p_{18} & p_{19} & p_{20} & p_{21} \end{array} \\ \left[\begin{array}{ccccccccccccccccccccc} 0&0&0&0&0&0&0&0&0&0&0&0&0&0&0&0&0&0&0&0&0 \\ 0&0&0&1&1&0&0&0&0&0&0&0&0&0&0&0&0&0&0&0&0 \\ 0&0&0&1&1&0&0&0&0&0&0&0&0&0&0&0&0&0&0&0&0 \\ 0&1&1&0&0&0&0&0&0&0&0&0&0&0&0&0&0&0&0&0&0 \\ 0&1&1&0&0&0&0&0&0&0&0&0&0&0&0&0&0&0&0&0&0 \\ 0&0&0&0&0&0&1&1&1&1&1&1&1&1&1&1&1&1&1&0&0 \\ 0&0&0&0&0&1&0&0&0&0&0&0&0&1&1&1&1&1&1&0&0 \\ 0&0&0&0&0&1&0&0&0&0&0&0&0&1&1&1&1&1&1&0&0 \\ 0&0&0&0&0&1&0&0&0&0&0&0&0&1&1&1&1&1&1&0&0 \\ 0&0&0&0&0&1&0&0&0&0&0&0&0&1&1&1&1&1&1&0&0 \\ 0&0&0&0&0&1&0&0&0&0&0&0&0&1&1&1&1&1&1&0&0 \\ 0&0&0&0&0&1&0&0&0&0&0&0&0&1&1&1&1&1&1&0&0 \\ 0&0&0&0&0&1&0&0&0&0&0&0&0&1&1&1&1&1&1&0&0 \\ 0&0&0&0&0&1&1&1&1&1&1&1&1&0&0&0&1&1&1&0&0 \\ 0&0&0&0&0&1&1&1&1&1&1&1&1&0&0&0&1&1&1&0&0 \\ 0&0&0&0&0&1&1&1&1&1&1&1&1&0&0&0&1&1&1&0&0 \\ 0&0&0&0&0&1&1&1&1&1&1&1&1&1&1&1&0&0&0&0&0 \\ 0&0&0&0&0&1&1&1&1&1&1&1&1&1&1&1&0&0&0&0&0 \\ 0&0&0&0&0&1&1&1&1&1&1&1&1&1&1&1&0&0&0&0&0 \\ 0&0&0&0&0&0&0&0&0&0&0&0&0&0&0&0&0&0&0&0&0 \\ 0&0&0&0&0&0&0&0&0&0&0&0&0&0&0&0&0&0&0&0&0 \end{array}\right] \end{array} \begin{array}{l} \\ p_1 \\ p_2 \\ p_3 \\ p_4 \\ p_5 \\ p_6 \\ p_7 \\ p_8 \\ p_9 \\ p_{10} \\ p_{11} \\ p_{12} \\ p_{13} \\ p_{14} \\ p_{15} \\ p_{16} \\ p_{17} \\ p_{18} \\ p_{19} \\ p_{20} \\ p_{21} \end{array}$$

Fig. 5.7 Neighborhood matrix of concurrency graph for PN_2

Fig. 5.8 Incidence matrix of a concurrency hypergraph for PN_1

$$A_{\mathcal{H}_{C_1}} = \begin{array}{c} \\ \\ \\ \end{array} \begin{array}{cccccc} p_1 & p_2 & p_3 & p_4 & p_5 & p_6 \end{array}$$

$$A_{\mathcal{H}_{C_1}} = \begin{bmatrix} 1 & 1 & 0 & 0 & 0 & 1 \\ 0 & 1 & 1 & 1 & 0 & 0 \\ 1 & 0 & 1 & 0 & 1 & 0 \end{bmatrix} \begin{matrix} M_0 \\ M_1 \\ M_2 \end{matrix}$$

A generalization of the concurrency graph is a *concurrency hypergraph*, initially introduced in [32] and formally defined in [31]. Similar to the graph, vertices of the concurrency hypergraph refer to the places of the Petri net. However, its hyperedges may by incident to more than two vertices.

Definition 5.5 (*Concurrency hypergraph*) *Concurrency hypergraph* $\mathcal{H}_C = (P, \mathcal{M})$ of an interpreted Petri net $PN = (P, T, F, M_0, X, Y)$ is a hypergraph, which vertices refer to the places of the net. The set of hyperedges corresponds to all reachable markings in the Petri net: $\mathcal{M} = \{M_0, \ldots, M_{k-1}\}$, where M_0 is an initial marking, and k is the number of all reachable markings in PN.

Clearly, concurrency hypergraph refers to the concurrency set (cf. Definition 2.18) of a Petri net. Therefore, it can be directly obtained from the reachability graph (cf. Definition 2.19) of the net.

Figure 5.8 illustrates the concurrency hypergraph for the net PN_1. There are six vertices that refer to the net places: $P = \{p_1, \ldots, p_6\}$. Three hyperedges directly correspond to all reachable markings: $\mathcal{M} = \{M_0, M_1, M_2\}$.

Let us now present a universal algorithm for obtaining the concurrency hypergraph for a Petri net. The method based on the typical idea of computation of all reachable markings in the Petri net, taken from [5].

Algorithm 5.1 Computation of the concurrency hypergraph for a Petri net

Input: Petri net $PN = (P, T, F, M_0)$
Output: Concurrency Hypergraph $\mathcal{H}_C = (P, \mathcal{M})$

$\mathcal{H} \leftarrow \{M_0\}$
$\mathcal{Q}$.push($\{M_0\}$)
while Q is not empty **do**
  $\{M\} = \mathcal{Q}$.pop()
  for all $t \in M[>$ **do**
    generate M' with $M[t>M'$
    if $M' \notin \mathcal{H}$ **then**
      $\mathcal{Q}$.push($\{M'\}$)
      $\mathcal{H} = \mathcal{H} \cup \{M'\}$
    end if
  end for
end while

Unfortunately, the number of reachable markings in the Petri net may be exponential with respect to the size of the net (expressed for example by the number of

places). Such a situation is called *state explosion problem* [27]. For example, the number of hyperedges in the concurrency hypergraph $\mathcal{H}_{C_2}$ of the net PN_2 is equal to $|\mathcal{M}| = 70$, which is more, than the number of edges in the corresponding concurrency graph ($|E| = 68$).
Let us illustrate Algorithm 5.1 by an example. Recall interpreted Petri net PN_3 from Fig. 2.6. The initial marking of the traffic light system involves the following places: $M_0 = \{p_1, p_4, p_6\}$. Such a state forms the first edge of the concurrency hypergraph $\mathcal{H}_{C_3}$ and is enqueued in $\mathcal{Q}$. At the first interaction of the *while...do* loop, M_0 is dequeued for the analysis. There are two transitions enabled in this state $M_0[t_1{>}M_1$ and $M_0[t_4{>}M_2$, where $M_1 = \{p_2, p_6\}$ and $M_2 = \{p_4, p_5\}$, respectively. Since both markings have not been included in the concurrency hypergraph, they are enqueued in $\mathcal{Q}$ and added to $\mathcal{H}_{C_3}$. At the second execution of the *while...do* loop, marking M_1 is dequeued. There is only one enabled transition at this state: $M_1[t_2{>}M_3$, where $M_3 = \{p_3, p_6\}$. Marking M_3 is enqueued in $\mathcal{Q}$ and added to $\mathcal{H}_{C_3}$. Further interactions of the *while...do* loop include analysis of enqueued markings: $\mathcal{Q} = \{M_2, M_3\}$, but no more hyperedges are added to $\mathcal{H}_{C_3}$. Therefore, the final concurrency hypergraph consists of four hyperedges, as it is shown in Fig. 5.9.

There are several techniques, that can be used to avoid the state explosion during formation of the reachability set. Let us briefly present the most popular ones. The idea of *net reduction* considers simplification of the initial concurrent system, by application of fusion of series (or parallel) of places (or transitions) [1, 24]. Such a method is especially valuable in the analysis of the behavior of a Petri net, since it preserves liveness and safeness of the initial system [2–4, 15, 24, 26, 35]. However, it is not directly oriented on the reachability set formation, but can be useful in conjunction with other techniques.

Other popular method of analysis bases on the generation of the *reduced reachability graph* [2–4, 6, 14–16, 24, 26, 29, 30, 35]. The aim of this technique is to reduce the time and the space during formation of the reachability graph [12]. Unfortunately, very often formation of the reduced graph does not preserve the full information about the concurrency relation between particular places [15]. Moreover, the size of the reduced graph can be still to large to be computed [12].

The above techniques (reduction of the initial net, formation of the reduced reachability graph) are not considered in this book. However, in some cases they can be successfully applied to the presented analysis and decomposition methods of concurrent systems [1, 15, 31, 32, 34, 35]

Fig. 5.9 Incidence matrix of a concurrency hypergraph for PN_1

$$A_{\mathcal{H}_{C_3}} = \begin{array}{c} \begin{array}{cccccc} p_1 & p_2 & p_3 & p_4 & p_5 & p_6 \end{array} \\ \left[\begin{array}{cccccc} 1 & 0 & 0 & 1 & 0 & 1 \\ 0 & 1 & 0 & 0 & 0 & 1 \\ 0 & 0 & 0 & 1 & 1 & 0 \\ 0 & 0 & 1 & 0 & 0 & 1 \end{array}\right] \end{array} \begin{array}{c} \\ M_0 \\ M_1 \\ M_2 \\ M_3 \end{array}$$

Instead, we will present other method of concurrency (and further sequentiality) analysis. The idea relies on the *structural concurrency relation* in the net and was initially proposed in [17] with further enhancement in [18].

Definition 5.6 (*Structural concurrency relation*) *Structural concurrency relation* $||^A \subseteq P \times P$ on the set of places P of an interpreted Petri net $PN = (P, T, F, M_0, X, Y)$ is the smallest symmetric relation such that

1. $\forall p, p' \in P : (p \in M_0 \wedge p' \in M_0) \Rightarrow (p, p') \in ||^A$;
2. $\forall t \in T\ \forall (p, p') \in t\bullet : p \neq p' \Rightarrow (p, p') \in ||^A$;
3. $\forall p \in P\ \forall t \in T : ((\forall p' \in \bullet t : (p, p') \in ||^A) \Rightarrow (\forall p'' \in t\bullet : (p, p'') \in ||^A))$.

The three rules presented above permit to obtain a structural concurrency relation of the net. The first one simply checks the initial state, since any two places marked at M_0 are structurally concurrent. The second condition states, that transition output places are structurally concurrent. The third rule implies that if all the input places of the transition are structurally concurrent to a particular place, then all the output places of such a transition are structurally concurrent to this place.

Algorithm 5.2 presents the idea of computation of $||^A$ of a Petri net. The computational complexity of the algorithm is bounded by $O(x^5)$, where x is the number of places and transitions of the net [17]. Further estimations and improvements of the algorithm have shown, that such a complexity can be even decreased for EFC-nets to $O(x^3)$ [18].

Algorithm 5.2 Computation of the structural concurrency relation $||^A$

Input: Petri net $PN = (P, T, F, M_0)$
Output: Concurrency relation $||^A \subseteq P \times P$

```
||^A = {(p, p') : (p ∈ M_0 ∧ p' ∈ M_0)} ∪ ⋃_{t∈T} t• × t•
repeat
    R = ||^A;
    T = T
    while T ≠ ∅ do
        choose t ∈ T
        P := {p' ∈ P : ∀p ∈ •t (p, p') ∈ ||^A}
        ||^A := ||^A ∪ {(p, p'), (p', p) : p ∈ t•, p' ∈ P}
        T = T \ {t}
    end while
until R = ||^A
```

Unfortunately, the structural concurrency relation does not always coincide with real concurrency relation. Consider the net presented in Fig. 5.10 (a), taken from [17]. Its concurrency graph is shown in Fig. 5.10 (b), while $||^A$ relation is illustrated in Fig. 5.10 (c). Clearly, $||^A$ contains redundant pairs of places that are not concurrent. Particulary, $\{(p_1, p_4), (p_2, p_4), (p_3, p_4)\} \in ||^A$, while none of those are in concurrency relation $||$. Thus, for this net we have $|| \neq ||^A$. Moving on, there are two important theorems that state about coincidence of relations $||$ and $||^A$:

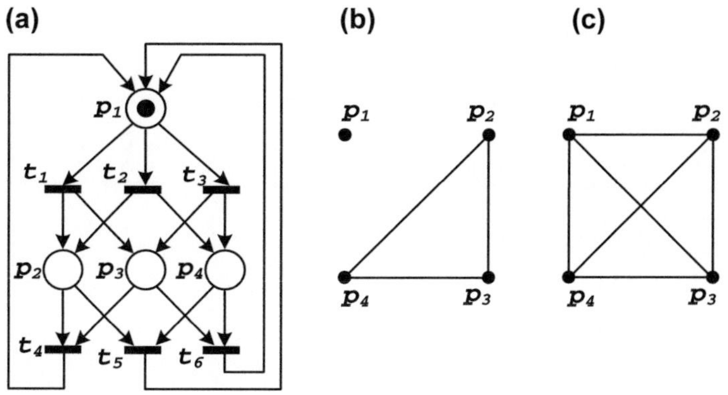

Fig. 5.10 Example of a net (taken from [17]) where $||^A$ does not coincide with $||$

Theorem 5.1 *[18] For every net system,* $|| \subseteq ||^A$. □

Theorem 5.2 *[18] The relations* $||$ *and* $||^A$ *coincide for live and bounded EFC-nets.*

Since every interpreted Petri net is live and safe (and therefore bounded), we have immediately the following:

Theorem 5.3 *The relations* $||$ *and* $||^A$ *coincide for interpreted EFC-nets.*

Proof Follows directly from Theorem 5.2. □

5.3 Sequentiality Analysis

Based on Definition 2.17, two places are *sequential* if there is no state where both places are marked simultaneously. Similarly to the *concurrency*, such a property can represent by an undirected graph.

Definition 5.7 (*Sequentiality graph*) *Sequentiality graph* $G_S = (P, \mathcal{E})$ of an interpreted Petri net $PN = (P, T, F, M_0, X, Y)$ is a graph, which vertices refer to the places of the net. Two vertices are connected by an edge if corresponding places are sequential in PN, i.e., they are not simultaneously marked in any reachable marking.

Sequentiality graph G_{S_1} for PN_1 and its neighborhood matrix is shown in Fig. 5.11. Six edges illustrates the sequentiality relation in the net. Since concurrency and sequentiality relation are complementary, graph G_{S_1} is complementary to G_{C_1} and vice versa. Let us formally define such a property.

Definition 5.8 (*Complement of a concurrency/sequentiality graph*) Let $G_C = (P, E)$ be a concurrency graph of an interpreted Petri net $PN = (P, T, F, M_0, X, Y)$, and let $G_S = (P, \mathcal{E})$ be a sequentiality graph of PN. Then, $\overline{G_C} = G_S$ and $\overline{E} = \mathcal{E}$. Similarly, $\overline{G_S} = G_C$ and $\overline{\mathcal{E}} = E$.

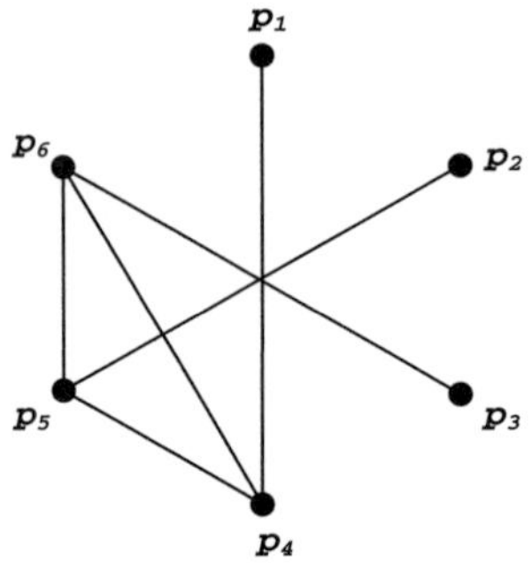

$$N_{G_{S_1}} = \begin{array}{c} \begin{array}{cccccc} p_1 & p_2 & p_3 & p_4 & p_5 & p_6 \end{array} \\ \begin{bmatrix} 0 & 0 & 0 & 1 & 0 & 0 \\ 0 & 0 & 0 & 0 & 1 & 0 \\ 0 & 0 & 0 & 0 & 0 & 1 \\ 1 & 0 & 0 & 0 & 1 & 1 \\ 0 & 1 & 0 & 1 & 0 & 1 \\ 0 & 0 & 1 & 1 & 1 & 0 \end{bmatrix} \end{array} \begin{array}{c} \\ p_1 \\ p_2 \\ p_3 \\ p_4 \\ p_5 \\ p_6 \end{array}$$

Fig. 5.11 Sequentiality graph for PN_1 and its neighborhood matrix

Definition 5.9 (*Structural sequentiality relation*) Structural sequentiality relation of an interpreted Petri net $PN = (P, T, F, M_0, X, Y)$ is the smallest symmetric relation such that $\forall p, p' \in P : (p, p') \notin ||^A \Rightarrow (p, p')$ are in the structural sequentiality relation (are structurally sequential).

Recall Theorem 5.1. It states, that for every net system, $|| \subseteq ||^A$. That is, the real concurrency relation is a subset of the structural one. Since concurrency and sequentiality relations are complementary, we can formulate a very important property

Theorem 5.4 *For every net system, the structural sequentiality relation is a subset of the real sequentiality relation.* □

Proof Follows from Theorem 5.1 and Definition 2.15. □

Note, that sequentiality relation is closely related to *state machine components*. Each place of the particular component is sequential to the other places that belong to such SMC. Let us formally define the set of all SMCs as a *sequentiality hypergraph*. Such a structure was initially proposed in [31], now we will define it more formally.

Definition 5.10 (*Sequentiality hypergraph*) *Sequentiality hypergraph* $\mathcal{H}_S = (P, \mathcal{S})$ of an interpreted Petri net $PN = (P, T, F, M_0, X, Y)$ is a hypergraph, which vertices refer to the places of the net. The set of hyperedges directly corresponds to all state machine components in the Petri net: $\mathcal{S} = \{S_1, \ldots, S_l\}$, where l is the number of all SMCs in PN.

An incidence matrix for the sequentiality hypergraph $\mathcal{H}_{S_1}$ is shown in Fig. 5.12. Each hyperedge refer to an SMC, thus there are four state machine components in PN_1.

Furthermore, Fig. 5.13 illustrates a matrix of sequentiality hypergraph $\mathcal{H}_{S_2}$ for PN_2. Please note, that there are only $|\mathcal{S}| = 8$ edges in $\mathcal{H}_{S_2}$, since concurrency hypergraph for this net contains 70 hyperedges.

$$A_{\mathcal{H}_{S_1}} = \begin{array}{c} \begin{matrix} p_1 & p_2 & p_3 & p_4 & p_5 & p_6 \end{matrix} \\ \begin{bmatrix} 1 & 0 & 0 & 1 & 0 & 0 \\ 0 & 1 & 0 & 0 & 1 & 0 \\ 0 & 0 & 1 & 0 & 0 & 1 \\ 0 & 0 & 0 & 1 & 1 & 1 \end{bmatrix} \end{array} \begin{matrix} S_1 \\ S_2 \\ S_3 \\ S_4 \end{matrix}$$

Fig. 5.12 Incidence matrix of a sequentiality hypergraph for PN_1

$$A_{\mathcal{H}_{S_2}} = \begin{array}{c} \begin{matrix} p_1 & p_2 & p_3 & p_4 & p_5 & p_6 & p_7 & p_8 & p_9 & p_{10} & p_{11} & p_{12} & p_{13} & p_{14} & p_{15} & p_{16} & p_{17} & p_{18} & p_{19} & p_{20} & p_{21} \end{matrix} \\ \begin{bmatrix} 1 & 1 & 1 & 0 & 0 & 1 & 0 & 0 & 0 & 0 & 0 & 0 & 0 & 0 & 0 & 0 & 0 & 0 & 0 & 1 & 1 \\ 1 & 1 & 1 & 0 & 0 & 0 & 1 & 1 & 1 & 1 & 1 & 1 & 1 & 0 & 0 & 0 & 0 & 0 & 0 & 1 & 1 \\ 1 & 1 & 1 & 0 & 0 & 0 & 0 & 0 & 0 & 0 & 0 & 0 & 0 & 1 & 1 & 1 & 0 & 0 & 0 & 1 & 1 \\ 1 & 1 & 1 & 0 & 0 & 0 & 0 & 0 & 0 & 0 & 0 & 0 & 0 & 0 & 0 & 0 & 1 & 1 & 1 & 1 & 1 \\ 1 & 0 & 0 & 1 & 1 & 1 & 0 & 0 & 0 & 0 & 0 & 0 & 0 & 0 & 0 & 0 & 0 & 0 & 0 & 1 & 1 \\ 1 & 0 & 0 & 1 & 1 & 0 & 1 & 1 & 1 & 1 & 1 & 1 & 1 & 0 & 0 & 0 & 0 & 0 & 0 & 1 & 1 \\ 1 & 0 & 0 & 1 & 1 & 0 & 0 & 0 & 0 & 0 & 0 & 0 & 0 & 1 & 1 & 1 & 0 & 0 & 0 & 1 & 1 \\ 1 & 0 & 0 & 1 & 1 & 0 & 0 & 0 & 0 & 0 & 0 & 0 & 0 & 0 & 0 & 0 & 1 & 1 & 1 & 1 & 1 \end{bmatrix} \end{array} \begin{matrix} S_1 \\ S_2 \\ S_3 \\ S_4 \\ S_5 \\ S_6 \\ S_7 \\ S_8 \end{matrix}$$

Fig. 5.13 Incidence matrix of a sequentiality hypergraph for PN_2

Finally, let us introduce a very important association between support of a place invariant and a state machine component in a Petri net. Clearly, both sets are closely related. However, not all obtained p-invariants refer to proper SMCs. Invariants are calculated basing only on the structure of the net, thus markings are not taken into account. For example, invariant $\overrightarrow{y_4}$ in net PN_1 supports places p_1 and p_2 (cf. Fig. 5.4). Both places are initially marked and they are concurrent. Therefore, such an invariant does not form a proper SMC.

Indeed, an SMC corresponds to the support of the p-invariant that contains a single token in the initial marking. Such a relation was theoretically proved in [10], while its practical application can bc found in [7, 8, 25]. Let us formulate it formally for the interpreted Petri nets.

Theorem 5.5 *A state machine component of an interpreted Petri net corresponds to the support of place invariant containing exactly one token in the initial marking.*

Proof Follows directly from Proposition 5.7 [10]. □

There are four place invariants in PN_1 (cf. Fig. 5.4) that contain exactly one token in the initial marking: $\overrightarrow{y_1}, \overrightarrow{y_2}, \overrightarrow{y_3}, \overrightarrow{y_5}$. According to the above theorem, their supports $I_1 = \{p_1, p_4\}$,$\{p_3, p_6\}$, $I_2 = \{p_2, p_5\}$, $I_3 = \{p_1, p_2, p_3\}$, $I_5 = \{p_4, p_5, p_6\}$ correspond to the SMCs in this net. Indeed, four hyperedges of $\mathcal{H}_{S_1}$ directly refer to the supports of the obtained invariants.

Furthermore, all eight supports of invariants of PN_2 contain exactly one token in the initial marking. Therefore, they correspond to the edges of $\mathcal{H}_{S_2}$.

5.4 Properties of Concurrency and Sequentiality Hypergraphs

This section presents properties and associations between concurrency and sequentiality hypergraphs of an interpreted Petri net.

At the beginning, let us introduce the essential theorem regarding exact transversals in a concurrency hypergraph and state machine components of an interpreted net. Similar property was initially proposed in [31], but restricted to MG-nets only. Let us extend it for any well-formed Petri net.

Theorem 5.6 *An exact transversal of the concurrency hypergraph of a well-formed Petri net directly refers to the set of places of a state machine component in this net.*

Proof Let $\mathcal{H}_C = (P, \mathcal{M})$ be a concurrency hypergraph of a well-formed Petri net $PN = (P, T, F, M_0)$, and let X be an exact transversal of $\mathcal{H}$.

To prove the whole theorem, we shall prove, that X satisfies all six conditions of the set of places that form a state machine component (according to Definition 2.25).

From the definition of an exact transversal we know, that X intersects each hyperedge of the concurrency hypergraph. Since $\mathcal{H}_C$ is a hypergraph on vertices P and hyperedges $\mathcal{M}$, the set X contains places of a Petri net that intersects each marking exactly once. It implies properties directly referring to the definition of state machine component

(a) all the places of X are sequential, since they are not marked simultaneously (1st condition of an SMC),
(b) $X \subseteq P$ (3rd condition),
(c) only one place of X is initially marked at M_0 (6th condition).

On the other hand, we know that *PN* is well-formed, thus it is live, safe and reversible. It means that each marking is reachable from any other marking. Thus, each place of X is reachable from any other place (since it intersects each marking exactly once). It finally implies the remaining conditions of SMC

(d) X is strongly connected (2nd condition),
(e) there exists a set of transitions $T' \in T$ that permits to obtain each reachable marking (4th condition),
(f) there exists a set of arcs $F' \in F$ that connects all the places X and transitions T' (and thus permits to obtain each reachable marking, 5th condition).

Since all the conditions are satisfied, exact transversal X directly refers to the set of places of an SMC. □

Since every interpreted Petri net is well formed, we have immediately the following:

Theorem 5.7 *An exact transversal in the concurrency hypergraph of an interpreted Petri net forms an edge in the sequentiality hypergraph.*

Proof Follows directly from Theorem 5.6 and Definition 2.15. □

Based on the above theorem we can easily observe the following association between concurrency and sequentiality hypergraphs:

Theorem 5.8 *An exact transversal in the concurrency hypergraph of an interpreted Petri net forms an edge in the sequentiality hypergraph.*

Proof Follows directly from Theorem 5.7 and Definition 5.10. □

There are four exact transversals in PN_1: $\mathcal{X} = \{X_1, \ldots, X_4\}$, where $X_1 = \{p_1, p_4\}$, $X_2 = \{p_2, p_5\}$, $X_3 = \{p_3, p_6\}$ and $X_1 = \{p_4, p_5, p_6\}$. They refer to the four SMCs, that can be obtained in such a net (cf. Fig. 2.8). Furthermore, all of achieved exact transversals directly correspond to hyperedges of the sequentiality hypergraph (cf. Fig. 5.12).
Assume, that the concurrency hypergraph of an interpreted Petri net is c-exact. Now we can notice very interesting properties and associations between concurrency and sequentiality hypergraphs

Theorem 5.9 *Let $\mathcal{H}_C$ be a c-exact concurrency hypergraph of an interpreted Petri net PN, and let $\mathcal{H}_S$ denote the sequentiality hypergraph of PN. Then, $\mathcal{H}_S$ is a compatibility hypergraph of $\mathcal{H}_C$: $\mathcal{H}_S = \mathcal{H}_C^{\mathcal{C}}$.*

Proof Follows directly from Theorems 4.2 and 5.8. □

Theorem 5.10 *Let $\mathcal{H}_C$ be a c-exact and conformal concurrency hypergraph of an interpreted Petri net PN, and let $\mathcal{H}_S$ denote the sequentiality hypergraph of PN. Then, $\mathcal{H}_C = (\mathcal{H}_C^{\mathcal{C}})^{\mathcal{C}} = (\mathcal{H}_S)^{\mathcal{C}}$.*

Proof Follows directly from Theorems 4.3 and 5.9. □

Theorem 5.11 *Let $\mathcal{H}_C$ be a c-exact and conformal concurrency hypergraph of an interpreted Petri net PN, and let $\mathcal{H}_S$ denote the sequentiality hypergraph of PN. Then $\mathcal{H}_S$ is also c-exact.*

Proof Follows directly from Theorems 4.4 and 5.9. □

Finally, we can state the following theorem:

Theorem 5.12 *Let $\mathcal{H}_S$ be a c-exact sequentiality hypergraph of an interpreted Petri net PN, and let $\mathcal{H}_C$ denote the concurrency hypergraph of PN. Then $\mathcal{H}_C = (\mathcal{H}_S)^{\mathcal{C}}$.*

Proof Follows directly from Theorems 4.4 and 5.9. □

Concurrency hypergraph $\mathcal{H}_{C_1}$ of PN_1 is c-exact. Therefore, the sequentiality hypergraph $\mathcal{H}_{S_1}$ of PN_1 is a compatibility hypergraph of $\mathcal{H}_{C_1}$, and can be obtained by calculating of all exact transversals in $\mathcal{H}_{C_1}$. However, $\mathcal{H}_{C_1}$ is not conformal, thus $\mathcal{H}_{S_1}$ is not c-exact.

Fig. 5.14 Incidence matrix of a sequentiality hypergraph for PN_3

$$A_{\mathcal{H}_{S_3}} = \begin{array}{c} \begin{matrix} p_1 & p_2 & p_3 & p_4 & p_5 & p_6 \end{matrix} \\ \begin{bmatrix} 0 & 1 & 1 & 1 & 0 & 0 \\ 0 & 0 & 0 & 0 & 1 & 1 \\ 1 & 1 & 1 & 0 & 1 & 0 \end{bmatrix} \end{array} \begin{matrix} S_1 \\ S_2 \\ S_3 \end{matrix}$$

Concurrency hypergraph $\mathcal{H}_{C_2}$ of PN_2 is c-exact. It means, that the sequentiality hypergraph $\mathcal{H}_{S_2}$ of PN_2 can be directly obtained by computation of all exact transversals in $\mathcal{H}_{C_2}$. Additionally, $\mathcal{H}_{S_2}$ is conformal. Therefore, the compatibility of the sequentiality hypergraph results in the initial concurrency hypergraph $\mathcal{H}_{C_2}$.

Let us point out the importance of Theorem 5.12. The presented property is especially valuable in the analysis of the interpreted Petri nets. It permits for the inverse computation of the concurrency set (and thus reachability set) directly from the sequentiality hypergraph. Let us explain it by an example.

As it was already shown, there are three invariants in the simplified traffic light system illustrated by PN_3. Their supports refer to the following sets of places: $I_1 = \{p_2, p_3, p_4\}$, $I_2 = \{p_5, p_6\}$, $I_3 = \{p_1, p_2, p_3, p_5\}$. All the above sets contain exactly one token in the initial marking $M_0 = \{p_1, p_4, p_6\}$. Therefore, they correspond to the state machine components in PN_3. Indeed, the sequentiality hypergraph $\mathcal{H}_{S_3}$ for this net contains three SMCs, as it is shown in Fig. 5.14.

Hypergraph $\mathcal{H}_{S_3}$ is conformal and c-exact. Therefore, we can obtain the concurrency hypergraph of PN_3 by calculation of all exact transversals in $\mathcal{H}_{S_3}$. There are four exact transversals in $\mathcal{H}_{S_3}$: $\mathcal{X} = \{X_1, X_2, X_3, X_4\}$, where $X_1 = \{p_1, p_4, p_6\}$, $X_2 = \{p_2, p_6\}$, $X_4 = \{p_4, p_5\}$ and $X_1 = \{p_3, p_6\}$, respectively. Notice, that each of the above sets directly corresponds to the reachable marking in PN_3 (cf. Fig. 5.9).

References

1. Banaszak Z, Kuś J, Adamski M (1993) Petri nets: modeling, control and synthesis of discrete systems. Higher School of Engineering Publishing House, Zielona Góra (in Polish)
2. Berthelot G (1986) Checking properties of nets using transformation. In: Advances in Petri Nets'85. Lecture notes in computer science, vol 222. Springer, pp 19–40
3. Berthelot G, Roucairol C (1976) Reduction of Petri nets. Mathematical foundations of computer science, vol 45. Lecture notes in computer science. Springer, Berlin, pp 202–209
4. Berthelot G, Roucairol C, Valk R (1980) Reduction of nets and parallel programs. In: Lecture Notes in computer science, vol. 84. Springer, pp 277–290
5. Buchholz P, Kemper P (2002) Hierarchical reachability graph generation for Petri nets. Formal Methods Syst Des 21(3):281–315
6. Clarke EM, Grumberg O, Minea M, Peled DA (1999) State space reduction using partial order techniques. STTT 2(3):279–287
7. Cortadella J (2002) Logic synthesis for asynchronous controllers and interfaces. Springer series in advanced microelectronics. Springer, Berlin, New York
8. Cortadella J, Kishinevsky M, Lavagno L, Yakovlev A (1998) Deriving Petri nets from finite transition systems. IEEE Trans Comput 47(8):859–882

9. Cortadella J, Reisig W (eds) 25th international conference applications and theory of Petri Nets 2004, ICATPN 2004, Bologna, Italy, 21–25 Jun 2004, Proceedings. Lecture notes in computer science, vol 3099. Springer
10. Desel J, Esparza J (1995) Free choice Petri nets. Cambridge University Press, New York, NY, USA
11. Dong M, Chen FF (2001) Process modeling and analysis of manufacturing supply chain networks using object-oriented Petri nets. Rob Comput Integr Manuf 17(1):121–129
12. Finkel A (1991) The minimal coverability graph for Petri nets. In: Advances in Petri Nets 1993, Papers from the 12th international conference on applications and theory of Petri nets, Gjern, Denmark, June 1991, pp 210–243
13. Holloway LE, Krogh BH (1990) Synthesis of feedback control logic for a class of controlled Petri nets. IEEE Trans Autom Control 35(5):514–523
14. Janicki R, Koutny M (1991) Using optimal simulations to reduce reachability graphs. In: Clarke EM, Kurshan RP (eds) Proceedings of the 2nd international conference on computer-aided verification CAV'90. LNCS, vol 531. Springer, London, pp 166–175
15. Karatkevich A (2007) Dynamic analysis of Petri net-based discrete systems. Lecture Notes in Control and Information Sciences, vol 356. Springer, Berlin
16. Klas G (1992) Hierarchical solution of generalized stochastic Petri nets by means of traffic processes. In: Jensen K (ed) Proceedingsof the 13th international conference on application and theory of Petri nets, Sheffield. Lecture notes in computer science, vol 616, pp 279–298
17. Kovalyov A (1992) Concurrency relation and the safety problem for Petri nets. In: Jensen K (ed) Proceedings of the 13th international conference on application and theory of Petri nets 1992. LNCS, vol 616. Springer, pp 299–309
18. Kovalyov A, Esparza J (1995) A polynomial algorithm to compute the concurrency relation of free-choice signal transition graphs. In: Proceedings of the international workshop WODES, pp 1–6
19. Lautenbach K (1986) Linear algebraic techniques for place/transition nets. In: Brauer W, Reisig W, Rozenberg G (eds) Advances in Petri nets. Lecture notes in computer science, vol 254. Springer, pp 142–167
20. Martinez J, Silva M (1982) A simple and fast algorithm to obtain all invariants of a generalized Petri net. In: Selected papers from the European workshop on applicational and theory of Petri nets, 1982. Springer, London, UK, pp 301–310
21. Memmi G, Roucairol G (1979) Linear algebra in net theory. In: Net theory and applications, proceedings of the advanced course on general net theory of processes and systems, Hamburg, 8–19 Oct 1979, pp 213–223
22. Montano L, García-Izquierdo, Villarroel J, Using the time Petri net formalism for specification, validation, and code generation in robot-control applications. Int J Rob Res 19:59–76
23. Moody J, Yamalidou K, Lemmon M, Antsaklis P (1994) Feedback control of Petri nets based on place invariants. In: Proceedings of the 33rd IEEE conference on decision and control, 1994, vol 3. IEEE, pp 3104–3109
24. Murata T (1989) Petri nets: properties, analysis and applications. Proc IEEE 77:548–580
25. Pastor E, Cortadella J (1998) Efficient encoding schemes for symbolic analysis of Petri nets. In: DATE'98, pp 790–795
26. Peterson JL (1981) Petri net theory and the modeling of systems. Prentice Hall PTR, Upper Saddle River, NJ, USA
27. Silva M (1993) Introducing Petri nets. In: Practice of Petri nets in manufacturing. Springer, Netherlands, pp 1–62
28. Silva M, Terue E, Colom JM (1998) Linear algebraic and linear programming techniques for the analysis of place/transition net systems. Springer, Berlin, Heidelberg, pp 309–373
29. Valmari A (1991) Stubborn sets for reduced state space generation. In: Advances in Petri nets 1990. Lecture notes in computer science, vol 483. Springer, Berlin, Germany, pp 491–515
30. Varpaaniemi K (1998) On the Stubborn Set Method in Reduced State Space Generation. Ph.D. thesis, Helsinki University of Technology, Department of Computer Science and Engineering

31. Wiśniewska M (2012) Application of hypergraphs in decomposition of discrete systems. Lecture notes in control and computer science, vol 23. University of Zielona Góra Press, Zielona Góra
32. Wiśniewska M, Adamski M (2006) Hypergraphs in the state space analysis of concurrent automata described by Petri nets. Meas Autom Monit 52(7 bis):62–64 (in Polish)
33. Wiśniewski R, Stefanowicz Ł, Bukowiec A, Lipiński J (2014) Theoretical aspects of Petri nets decomposition based on invariants and hypergraphs, Zhangjiajie, China. Lecture notes in electrical engineering, vol 308, pp 371–376
34. Yakovlev A, Gomes L, Lavagno L (2000) Hardware design and Petri nets. Springer
35. Zakrevskij A, Pottosin Y, Cheremisinova L (2009) Design of logical control devices. TUT Press, Moskov

Chapter 6
Decomposition of Concurrent Control Systems

This chapter deals with decomposition of concurrent control systems. The main aim of such a process is to split the control algorithm into sequential subsystems. Each of decomposed modules can be implemented separately, even with application of various devices. Furthermore, the functionality of the particular sequential subsystem can be reconfigured without touching the rest of the device. Three SM-decomposition methods are shown in the chapter. All of them refer to the concurrent control systems described by an interpreted Petri net. The first one is based on the classical linear algebra and place invariants computation. The second method uses perfect graph theory, while the remaining one applies calculation of exact transversals.

Each of the presented decomposition concepts has advantages and weak points. However, regardless of which method is used, each of them leads to the same goal—decomposition of a concurrent controller into sequential modules. Sometimes, particular method is unable to find the solution, for example due to the computational complexity. It is assumed, that alternative decomposition method should be used instead.

Notice, that according to Definition 2.26, each place of the initial net belongs to exactly one of the decomposed SMCs. If a place exists more than one SMC, it is replaced by a NOP. Furthermore, various SM-decompositions may exist for a single net. Thus, it is assumed that the main criterion of the presented methods is the minimal number of decomposed SMCs. In case of methods based on p-invariants and exact transversals the minimal number of SMCs is achieved during the additional selection process. We shall present a modified selection method taken from [21] (with further enhancement in [14–16, 20, 23]). Of course, one may easily modify the proposed algorithms and adjust selection process for own purposes by applying any of known algorithms thus solving the set covering problem [3–6, 8, 10, 17].

© Springer International Publishing Switzerland 2017

R. Wiśniewski, *Prototyping of Concurrent Control Systems Implemented in FPGA Devices*, Advances in Industrial Control,
DOI 10.1007/978-3-319-45811-3_6

6.1 SM-Decomposition Based on Place Invariants

According to Theorem 5.5, an SMC in the interpreted net corresponds to the support of p-invariant, that contains exactly one token in the initial marking. Let us apply such a property in SM-decomposition of a concurrent control system described by an interpreted Petri net. At the beginning, we shall present the idea of the decomposition. Next, we explain such a method by examples.

6.1.1 *The Idea of Method*

The decomposition of the interpreted Petri net $PN = (P, T, F, X, Y, M_0)$ into the set of SMCs $\mathcal{S} = \{S_1, \ldots, S_n\}$ is divided into the following steps:

1. *Calculation of the set of place invariants in the interpreted Petri net.* The set of p-invariants may be obtained by any known linear algebra technique. In our examples we shall use the algorithm that is presented in Chap. 5. Let us denote supports of the achieved set by $\mathcal{I} = \{I_1, \ldots, I_k\}$, where k is the number of all obtained p-invariants.
   Note, that if the algorithm is unable to obtain the set of invariants due to the time complexity, other decomposition method should be used instead.
2. *Formation of the set $\mathcal{S}$.* The set $\mathcal{S}$ is obtained directly from the set $\mathcal{I}$. Each support of the invariant $I \in \mathcal{I}$ is examined whether it contains exactly one token in the initial marking M_0. If so, it forms a proper SMC and is added to the set $\mathcal{S}$. Finally, $\mathcal{S} = \{S_1, \ldots, S_m\}$, where m is the number of all obtained SMCs.
3. *Verification if all the places of the net are covered by elements from $\mathcal{S}$.* Such a verification is necessary in order to check if PN can be decomposed with the use of p-invariants to avoid spurious solutions (cf. [12, 13]). If the place $p \in P$ is not covered by any $S \in \mathcal{S}$, the method stops execution. It means that the algorithm is unable to find the solution and a different decomposition method should be used.
4. *Selection of state machine components in $\mathcal{S}$.* This stage is divided into the following substeps [21]:

   (a) *Formation of the selection hypergraph $\mathcal{H}_L = (\mathcal{S}, P)$.* Vertices of $\mathcal{H}_L$ refer to the obtained state machine components, while its edges correspond to the net places.

   (b) *Cyclic reduction of dominated edges and dominating vertices in $\mathcal{H}_L$.* This stage can be divided into three sub-steps that are executed until no more elements or sets can be reduced [10]:

      - *Essential vertices*: If an edge contains only one vertex, it is an essential vertex. Essential vertices ought to be a part of the final solution of selection process.

- *Reduction of dominating hyperedges*: An edge can be reduced (removed) from $\mathcal{H}_L$ if it contains (or is equal to) another hyperedge. Simply, p_i is reduced by p_j if $p_i \supseteq p_j$.
- *Reduction of dominated vertices*: If a vertex is incident to the same edges as another vertex, it does not influence the final result and can be reduced from $\mathcal{H}_L$. In other words, S_i is reduced by S_j if $\forall p \in P: S_j \in p \Rightarrow S_i \in p$.
- *Reduction of empty edges and isolated vertices*: If after the reduction of dominated vertices or dominating edges an edge remains empty, it can be removed from the hypergraph. Similarly, isolated vertices are removed as well.

The above reduction technique is taken from [10], where the idea of *cyclic reduction* was used in Boolean logic formula minimization. It was an integral part of *Espresso* system [11], but has wide application fields, for example, as a part of Quine–McCluskey algorithm [2]. The main benefit of the presented method is its computational complexity. The whole process is bounded by a polynomial in the number of elements and sets [10].

(c) *Computation of the minimal transversal T in $\mathcal{H}_L$*. Obtained transversal indicate elements of $\mathcal{S}$ that ought to be selected.

(d) *Remove SMCs from $\mathcal{S}$ that are not indicated by T*. After removing of the redundant SMCs, set $\mathcal{S}$ contains only selected components. Therefore, the selection process is finished.

5. *Replacement of repeated places by NOPs.* Finally, places (or set of places) that exist in more than one SMC are replaced by nonoperational ones. Such a modification is necessary in order to prevent proper functionality of the controller and prohibit execution of the same action by various components. It is assumed, that replacement of repeated places is up to the designer. However, it can be easily done automatically, based on the preset criteria. In our considerations we just use the simplest method by replacing places according to the lexicographical order (straight or reverse) of SMCs.

Let us analyze the computational complexity of the above algorithm. Clearly, the whole procedure is exponential. The main bottleneck is tied to the computation of place invariants. As it was shown in Sect. 6.1 the number of invariants may be exponent. Another computational weakness of the above method can be found in selection process. Finding of the minimal transversal in some cases may also be exponential. However, application of the cyclic reduction technique reduces the initial hypergraph, so the minimal covering can be found relatively easily [10]. Furthermore, if the selection hypergraph is a c-exact hypergraph (or even an xt-hypergraph), the whole selection process turns out to be polynomial (cf. [15, 16, 23]). In such a case, the first exact transversal is searched (since it is minimal one, as well).

6.1.2 Examples

Let us now explain the idea of SM-decomposition based on the p-invariants by examples. We shall use the nets presented in Chap. 2. At the beginning relatively small nets PN_1 and PN_3 are decomposed. Then, we will move to the more complicated system of milling process specified by PN_2. Note, that we use already achieved results for place invariants computation shown in Chap. 5.

Let us decompose the net PN_1 from Fig. 2.1. There are five p-invariants in PN_1 with the support: $I_1 = \{p_1, p_4\}$, $I_2 = \{p_3, p_6\}$, $I_3 = \{p_2, p_5\}$, $I_4 = \{p_1, p_2, p_3\}$, $I_5 = \{p_4, p_5, p_6\}$. Four of them form a proper SMC, while I_4 does not, since places p_1 and p_2 are both initially marked. Therefore, $\mathcal{S} = \{S_1, \ldots, S_4\}$, where:

- $S_1 = I_1 = \{p_1, p_4\}$,
- $S_2 = I_2 = \{p_3, p_6\}$,
- $S_3 = I_3 = \{p_2, p_5\}$,
- $S_4 = I_5 = \{p_4, p_5, p_6\}$.

All six places of PN_1 exist in achieved SMCs, therefore this net can be decomposed with application of p-invariants.

Next, the selection process ought to be executed. Selection hypergraph $\mathcal{H}_{L_1} = (\mathcal{S}, P)$ consists of six vertices that represent the obtained SMCs and four hyperedges that correspond to the places of the net. Figure 6.1 (left) shows an incidence matrix of $\mathcal{H}_{L_1}$. Clearly, columns of the matrix correspond to the SMCs, while rows refer to places of PN_1.

In the subsequent step, a *cyclic reduction* is applied. Let us analyze it step-by-step. There are three essential vertices in $\mathcal{H}_{L_1}$: S_1, S_2, S_3. They cannot be reduced and they ought to be a part of the final solution. Reduction of dominating hyperedges removes p_4, p_5, and p_6 from the hypergraph, since $p_4 \supseteq p_1$, $p_5 \supseteq p_2$, and $p_6 \supseteq p_3$, respectively. Further reduction of dominated vertices remains $\mathcal{H}_{L_1}$ untouched. Finally, place S_4 is isolated, thus it is removed from the $\mathcal{H}_{L_1}$.

No more places, nor edges are reduced in further execution of the cyclic reduction. Figure 6.1 (right) illustrates the final result of this operation. There is only one transversal in the reduced $\mathcal{H}_{L_1}$ and it consists of three elements: $T = \{S_1, S_2, S_3\}$. It means, that those three SMCs are the result of the selection process.

Finally, places that exist in more than one SMC ought to be replaced by NOPs. Since each place belongs to exactly one $S \in \mathcal{S}$, there is no need to apply nonoperational places. Eventually, PN_1 is decomposed into three SMCs: $\mathcal{S} = \{S_1, S_2, S_3\}$,

Fig. 6.1 Incidence matrix of $\mathcal{H}_{L_1}$ before (*left*) and after cyclic reduction (*right*)

$$A_{\mathcal{H}_{L_1}} = \begin{array}{c} \begin{matrix} S_1 & S_2 & S_3 & S_4 \end{matrix} \\ \begin{bmatrix} 1 & 0 & 0 & 0 \\ 0 & 1 & 0 & 0 \\ 0 & 0 & 1 & 0 \\ 1 & 0 & 0 & 1 \\ 0 & 1 & 0 & 1 \\ 0 & 0 & 1 & 1 \end{bmatrix} \end{array} \begin{matrix} p_1 \\ p_2 \\ p_3 \\ p_4 \\ p_5 \\ p_6 \end{matrix} \qquad A_{\mathcal{H}_{L_1}} = \begin{array}{c} \begin{matrix} S_1 & S_2 & S_3 \end{matrix} \\ \begin{bmatrix} 1 & 0 & 0 \\ 0 & 1 & 0 \\ 0 & 0 & 1 \end{bmatrix} \end{array} \begin{matrix} p_1 \\ p_2 \\ p_3 \end{matrix}$$

Fig. 6.2 Incidence matrix of $\mathcal{H}_{L_3}$ before (*left*) and after cyclic reduction (*right*)

$$A_{\mathcal{H}_{L_3}} = \begin{bmatrix} 0 & 0 & 1 \\ 1 & 0 & 1 \\ 1 & 0 & 1 \\ 1 & 0 & 0 \\ 0 & 1 & 1 \\ 0 & 1 & 0 \end{bmatrix} \begin{matrix} p_1 \\ p_2 \\ p_3 \\ p_4 \\ p_5 \\ p_6 \end{matrix} \quad (\text{columns } S_1\ S_2\ S_3) \qquad A_{\mathcal{H}_{L_3}} = \begin{bmatrix} 0 & 0 & 1 \\ 1 & 0 & 0 \\ 0 & 1 & 0 \end{bmatrix} \begin{matrix} p_1 \\ p_4 \\ p_6 \end{matrix} \quad (\text{columns } S_1\ S_2\ S_3)$$

where $S_1 = \{p_1, p_4\}$, $S_2 = \{p_3, p_6\}$, $S_3 = \{p_2, p_5\}$. Achieved components are graphically illustrated in Fig. 2.8a–c.

Let us now decompose a simplified traffic light system specified by the Petri net PN_3 shown in Fig. 2.6. Three place invariants can be obtained in the net. Each of them refers to a proper SMC, since their supports contain exactly one token in the initial marking

- $S_1 = I_1 = \{p_2, p_3, p_4\}$,
- $S_2 = I_2 = \{p_5, p_6\}$,
- $S_3 = I_3 = \{p_1, p_2, p_3, p_5\}$.

Each of six places of PN_3 exists in at least one $S \in \mathcal{S}$, thus the net can be decomposed with the use of place invariants.

Clearly, the selection hypergraph $\mathcal{H}_{L_3}$ consists of $|\mathcal{S}| = 3$ vertices and $|P| = 6$ edges, as it is shown in Fig. 6.2 (left). Note, that all three vertices are essential (due to hyperedges p_1, p_4 and p_6). Thus, the final solution must contain all the obtained SMCs. Indeed, the minimal transversal contains three elements in the reduced hypergraph shown in Fig. 6.2 (right).

Finally, nonoperational places ought to be applied in order to exchange places that exist in more than one SMCs. Let us use lexicographic order. That is, the first SMC remains untouched. If a place exists in the subsequent components it is replaced by a NOP. Furthermore, the second SMC is analyzed. Similarly, if a place that exists in such an SMC also belongs the subsequent SMCs, it is exchanged by a nonoperational place.

In the presented example, S_1 and S_2 remain unchanged, because set S_2 does not contain any elements from S_1. However, in case of S_3, three places ought to be replaced: p_2, p_3 (exist in S_1), and p_5 (exists in S_2). Note, that p_2 and p_3 can be replaced by a common NOP. Therefore, the final set $\mathcal{S}$ of decomposed SMCs is as follows:

- $S_1 = \{p_2, p_3, p_4\}$,
- $S_2 = \{p_5, p_6\}$,
- $S_3 = \{p_1, NOP_1, NOP_2\}$.

The decomposed net is shown in Fig. 6.3. The first component (a) controls the lights for cars. The second one (b) is in charge of turning lights for pedestrians. The third SMC (c) can be seen as a semaphore between the lights for cars and for

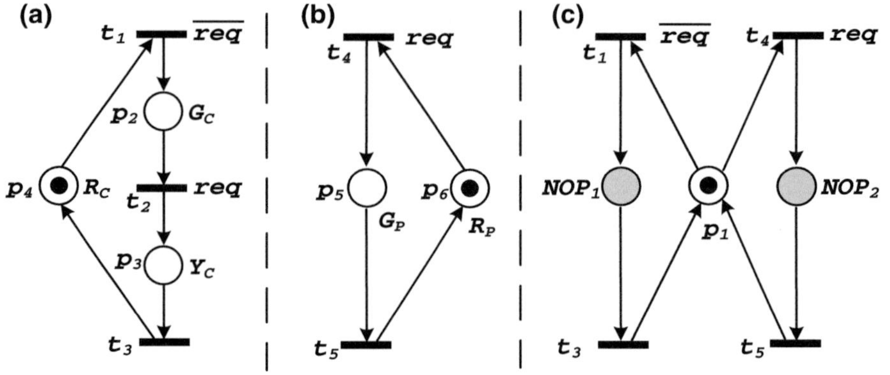

Fig. 6.3 Decomposed net PN_3 (method based on the place invariants computation)

pedestrians. It assures proper functionality of the whole system and ensures that the crossroad is collision-free.

Let us now analyze more complicated example. We shall decompose the milling process, specified by PN_2 (cf. Fig. 2.4). Eight invariants can be obtained in the net. All of them are marked exactly once in M_0, thus the set $\mathcal{S}$ of SMCs consists of eight elements as follows:

- $S_1 = \{p_1, p_2, p_3, p_6, p_{20}, p_{21}\}$,
- $S_2 = \{p_1, p_4, p_5, p_6, p_{20}, p_{21}\}$,
- $S_3 = \{p_1, p_2, p_3, p_7, p_8, p_9, p_{10}, p_{11}, p_{12}, p_{13}, p_{20}, p_{21}\}$,
- $S_4 = \{p_1, p_4, p_5, p_7, p_8, p_9, p_{10}, p_{11}, p_{12}, p_{13}, p_{20}, p_{21}\}$,
- $S_5 = \{p_1, p_2, p_3, p_{14}, p_{15}, p_{16}, p_{20}, p_{21}\}$,
- $S_6 = \{p_1, p_4, p_5, p_{14}, p_{15}, p_{16}, p_{20}, p_{21}\}$,
- $S_7 = \{p_1, p_2, p_3, p_{17}, p_{18}, p_{19}, p_{20}, p_{21}\}$,
- $S_8 = \{p_1, p_4, p_5, p_{17}, p_{18}, p_{19}, p_{20}, p_{21}\}$.

Clearly, all places of PN_2 exist in SMCs, therefore the net can be decomposed with the linear algebra technique. A selection hypergraph $\mathcal{H}_{L_2}$ for the achieved components consists of $|\mathcal{S}| = 8$ vertices and $|P| = 21$ hyperedges, as it is shown in Fig. 6.4 (left). There are no essential vertices in $\mathcal{H}_{L_2}$. Reduction of dominated hyperedges removes fourteen edges from the hypergraph:

- p_1 (because $p_1 \supseteq p_2$),
- p_3 (because $p_3 \supseteq p_2$),
- p_5 (because $p_5 \supseteq p_4$),
- p_8 (because $p_8 \supseteq p_7$),
- p_9 (because $p_8 \supseteq p_7$),
- p_{10} (because $p_{10} \supseteq p_7$),
- p_{11} (because $p_{11} \supseteq p_7$),
- p_{12} (because $p_{12} \supseteq p_7$),
- p_{13} (because $p_{13} \supseteq p_7$),

	S_1	S_2	S_3	S_4	S_5	S_6	S_7	S_8
p_1	1	1	1	1	1	1	1	1
p_2	1	0	1	0	1	0	1	0
p_3	1	0	1	0	1	0	1	0
p_4	0	1	0	1	0	1	0	1
p_5	0	1	0	1	0	1	0	1
p_6	1	1	0	0	0	0	0	0
p_7	0	0	1	1	0	0	0	0
p_8	0	0	1	1	0	0	0	0
p_9	0	0	1	1	0	0	0	0
p_{10}	0	0	1	1	0	0	0	0
p_{11}	0	0	1	1	0	0	0	0
p_{12}	0	0	1	1	0	0	0	0
p_{13}	0	0	1	1	0	0	0	0
p_{14}	0	0	0	0	1	1	0	0
p_{15}	0	0	0	0	1	1	0	0
p_{16}	0	0	0	0	1	1	0	0
p_{17}	0	0	0	0	0	0	1	1
p_{18}	0	0	0	0	0	0	1	1
p_{19}	0	0	0	0	0	0	1	1
p_{20}	1	1	1	1	1	1	1	1
p_{21}	1	1	1	1	1	1	1	1

	S_1	S_2	S_3	S_4	S_5	S_6	S_7	S_8
p_2	1	0	1	0	1	0	1	0
p_4	0	1	0	1	0	1	0	1
p_6	1	1	0	0	0	0	0	0
p_7	0	0	1	1	0	0	0	0
p_{14}	0	0	0	0	1	1	0	0
p_{17}	0	0	0	0	0	0	1	1

Fig. 6.4 Incidence matrix of $\mathcal{H}_{L_2}$ before (*left*) and after cyclic reduction (*right*)

- p_{15} (because $p_{15} \supseteq p_{14}$),
- p_{16} (because $p_{16} \supseteq p_{14}$),
- p_{18} (because $p_{18} \supseteq p_{17}$),
- p_{19} (because $p_{19} \supseteq p_{17}$),
- p_{20} (because $p_{20} \supseteq p_2$).
- p_{21} (because $p_{21} \supseteq p_2$).

No further cyclic reduction can be applied to $\mathcal{H}_{L_2}$. The final result of the cyclic reduction is presented in Fig. 6.4 (right). There are only six hyperedges in the reduced hypergraph, since fourteen edges have been removed.

The final solution of the selection process is obtained by computation of the minimal transversal in the reduced hypergraph. Notice, that there are more than one minimal transversal. Each of them refers to the minimal covering of the reduced $\mathcal{H}_{L_2}$. In our consideration, we shall include a transversal composed by the following SMCs: $T = \{S_1, S_3, S_6, S_7\}$.

Finally, repeated places ought to be replaced by NOPs. This time we use a reverse lexicographic order for choosing the particular SMCs. That is, the set S_7 remains unchanged, while replaced places are subsequently removed from components S_6, S_3, and eventually from S_1. The decomposed net is shown in Fig. 6.5.

The final set of decomposed sub-nets contains the following components:

- $S_7 = \{p_1, p_2, p_3, p_{17}, p_{18}, p_{19}, p_{20}, p_{21}\}$,

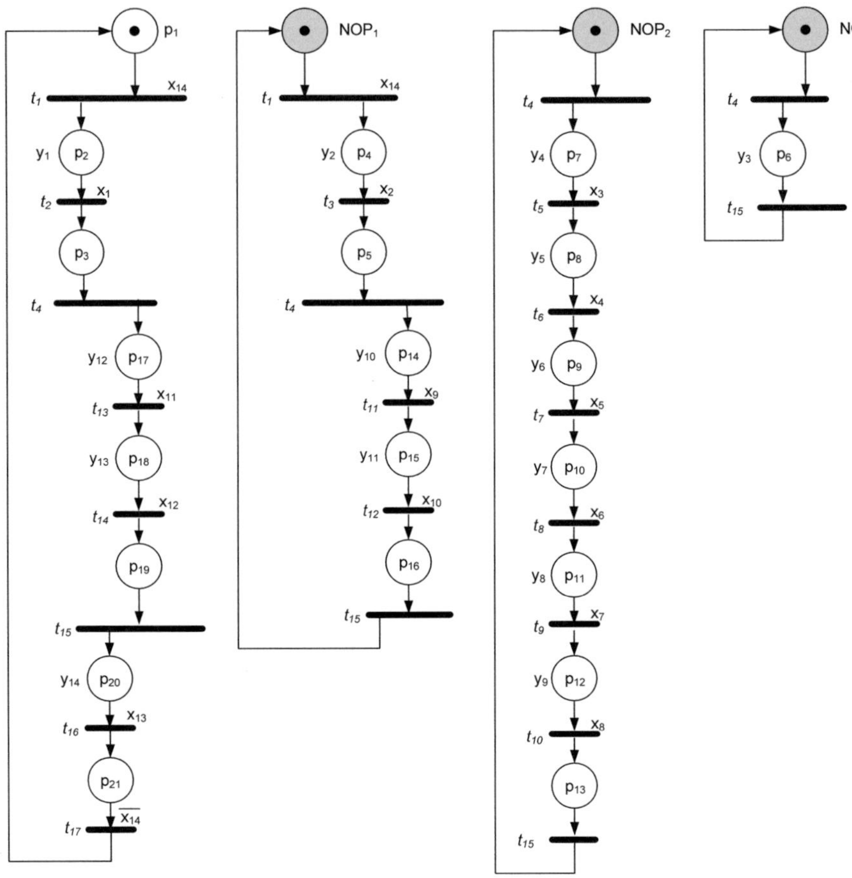

Fig. 6.5 Decomposed Petri net PN_2

- $S_6 = \{NOP_1, p_4, p_5, p_{14}, p_{15}, p_{16}\}$,
- $S_3 = \{NOP_2, p_7, p_8, p_9, p_{10}, p_{11}, p_{12}, p_{13}\}$,
- $S_1 = \{NOP_3, p_6\}$.

In the above SMCs, particular NOPs substitute the following places:

- $NOP_1 = \{p_1, p_{20}, p_{21}\}$,
- $NOP_2 = \{p_1, p_2, p_3, p_{20}, p_{21}\}$,
- $NOP_3 = \{p_1, p_2, p_3, p_{20}, p_{21}\}$.

Notice, that all three NOPs are initially marked. It is caused by place P_1, which is substituted in all of components, except S_7.

Let us now analyze achieved components. The first one (S_7) controls the movement of the wooden plank (signal y_1) and further drilling of assembling holes (y_{12}) on the one side of the plank. The holes on the reverse side of the plank are drilled (y_{10}) by the process executed by the second component (S_6). Additionally, such an SMC is in charge of setting up the main drill into the adequate position (sensor x_2). The process controlled by the third component (S_3) cuts out the proper shape from the wooden plank. Finally, the last SMC (S_1) simply controls a vacuum cleaner (y_3).

Note, that the above diversification of the initial net permits for easy and comfortable modification of the functionality of the prototyped Petri net. For example, if there is a need to exchange the shape that is cut out from the wooden plank, only component S_3 ought to be modified. Furthermore, there is a possibility to replace such a component with another, where different functionality is specified (for example cutting of the different shape). We shall show such a possibility in Chap. 9.

6.2 SM-Decomposition Based on Graph Theory

This section presents an innovate decomposition method, where comparability graphs are applied. The algorithm bases on the formation of the concurrency graph according to the structural concurrency relation. Next, if the graph is a comparability graph, it is colored in order to obtain decomposed state machine components.

The idea was initially proposed in [22]. We shall extend the method by simultaneous orientation and coloring of the concurrency graph. Furthermore, the idea of addition of nonoperational places is introduced.

Note, that presented decomposition method applies algorithms already shown in this book (especially referring to perfect graphs and structural concurrency relation). We also introduce an algorithm to supplement the achieved components by NOPs.

6.2.1 *The Idea of Method*

The decomposition of the interpreted Petri net $PN = (P, T, F, X, Y, M_0)$ into the set of SMCs $\mathcal{S} = \{S_1, \ldots, S_n\}$ with the application of comparability graphs is divided into the following steps:

1. *Computation of the structural concurrency relation* $||^A$ *for* PN. This step is executed according to Algorithm 5.2 presented in Chap. 5.
2. *Formation of the concurrency graph* G_C. Based on the achieved structural concurrency relation, the concurrency graph G_C is formed. Simply, vertices of G_C refer to the net places, while edges correspond to the obtained relation $||^A$. That is, two vertices of G_C are connected by an edge if they are structurally concurrent. Recall Theorem 5.4. It states, that for any net, a structural sequentiality relation is a subset of the real one. It means, that any set of places that are sequential in G_C, are also in a real sequential relation. However, in some cases, structural concurrency relation is not enough to perform the decomposition process based on the comparability graphs. We shall explain it further, in more details.

3. *Transitive orientation and coloring of* G_C. At this stage, G_C is transitively oriented and simultaneously coloring, according to Algorithm 3.2 presented in Chap. 3. It is an essential step of the whole decomposition process. If there exists a TRO in G_C, decomposition can be continued. Otherwise, a different decomposition method ought to be applied.
4. *Formation of the set* $\mathcal{S}$ *of decomposed components*. Clearly, each set of places obtained during coloring of G_C forms a sequential component S in $\mathcal{S}$. However, some of achieved components may be not strongly connected. If the set S contains a place with unconnected input or output transitions, it is supplemented by non-operational places, according to Algorithm 6.1. Simultaneously, the verification of achieved components is performed. The algorithm checks whether the set of unconnected input transitions and the set of unconnected output transitions are mutually exclusive empty (that is, one of the set is empty while the second is not empty). If so, the algorithm terminates execution, since the proposed decomposition method cannot be used. Otherwise, if all the components are successfully supplied by NOPs, the decomposition procedure finishes.

The achieved set $\mathcal{S}$ consists of decomposed state machine components. In opposite to the decomposition based on the place invariants, there is no need to perform additional selection process. If the structural concurrency relation coincides with real concurrency relation, the number of achieved components exactly composes the initial net.
Algorithm 6.1 supplies the achieved state machine components by non-operational places. First, each of obtained SMCs is examined if it is strongly connected. The algorithm searches for unconnected transition inputs and unconnected transition outputs. Clearly, if the net contains such transitions, it ought to be supplemented by a NOP. All the unconnected output transitions and all the unconnected input transitions of the particular component are tied to the same NOP.

Clearly, Algorithm 6.1 supplies only one nonoperational place to the single component. Therefore, the achieved results may be different than the SMCs obtained during the decomposition methods based on the place invariants or hypergraph theory, where multiple NOPs can be assigned to one component.
Let us analyze the computational complexity of the whole decomposition method. Computation of the structural concurrency relation and formation of the concurrency graph G_C is bounded by $O(x^5)$, where x is the number of places and transitions in the net [7] (cf. Chap. 5).

Simultaneous transitive orientation as well as coloring of G_C is performed in $O(|E| * \Delta(G_C))$ by Theorem 3.3. Since $E \subseteq P \times P$ and $\Delta(G_C) < n$, we shall enhance the upper bound to $O(n^3)$, where n is the number of places in the net ($n = |P|$).

Finally, execution of Algorithm 6.1 includes checking each of the achieved components. The total number of SMCs is equal to $|\mathcal{S}|$, thus the outer *for all* loop is executed at most n times, where n is the number of places in the net. The inner (*for all* loop) is bounded by $O(n * m)$, since it searches for unconnected input and

Algorithm 6.1 Supplementation of the components by NOPs

Input: An interpreted Petri net $PN = (P, T, F, M_0, X, Y)$, concurrency graph $G_C = (P, ||^A)$ of PN, set $\mathcal{S}$ obtained during coloring of G_C

Output: If PN is correctly decomposed: a set $\mathcal{S}$ supplemented by NOPs

```
for all S = (P', T', F', M'_0) ∈ 𝒮 do
    i = 1
    I ← ∅
    O ← ∅
    for all p ∈ P' do
        for each t ∈ •p such that •t = ∅ do I = I ∪ {t}
        for each t ∈ p• such that t• = ∅ do O = O ∪ {t}
    end for
    if [I ≠ ∅ or O ≠ ∅] then
        if [I = ∅ or O = ∅] then
            notify: "The net cannot be decomposed."
            return
        else
            P' = P' ∪ {NOP_i}
            for each t ∈ O do t• = {NOP_i}
            for each t ∈ I do •t = {NOP_i}
            if M'_0 = ∅ then M'_0 = {NOP_i}
            i ← i + 1
        end if
    end if
end for
```

output transitions. Thus, the computational complexity of Algorithm 6.1 is bounded by $O(n^2 * m)$.

Reassuming, the complexity of the whole decomposition method involves:

- Formation of the concurrency graph G_C, bounded by $O(x^5)$.
- Transitive orientation and coloring of G_C, bounded by $O(n^3)$.
- Verification and supplementation by NOPs, bounded by $O(n^2 * m)$.

Clearly, the computational complexity depends on the first step of the method. Therefore, we can finalize our analysis by the following statement: the computational complexity of the decomposition method based on the comparability graphs is bounded by $O(x^5)$, where x is the number of places and transitions in the net.

6.2.2 *Examples*

Let us now explain the idea of SM-decomposition based on the comparability graphs with examples. Similarly to the previous section, the nets presented in Chap. 2 will be used.

Let us start with the decomposition of PN_1. The concurrency graph G_{C_1} of such a net is shown in Fig. 6.6 (left). There is also an attempt of a transitive orientation

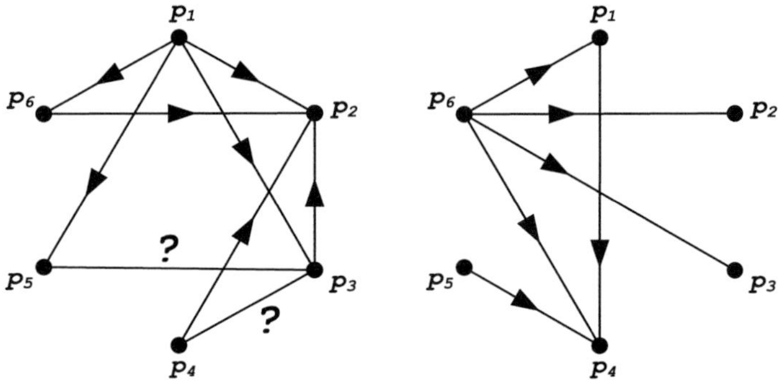

Fig. 6.6 An attempt concurrency graphs orientation for PN_1 (*left*) and PN_3 (*right*)

presented. Note, that edges $\{p_3, p_4\}$ and $\{p_3, p_5\}$ cannot be transitively oriented. Orientation $\overrightarrow{p_3p_4}$ forces direction of a missing edge $\overrightarrow{p_4p_5}$. On the other hand, orientation $\overrightarrow{p_4p_3}$ forces orientation $\overrightarrow{p_5p_3}$. However, $\overrightarrow{p_5p_3}$ forces the existence of a missing edge $\overrightarrow{p_5p_2}$. Therefore, G_{C_1} is not a comparability graph and cannot be decomposed with the proposed method.

Figure 6.6 (right) illustrates an attempt of transitive orientation of the concurrency graph G_{C_3} of the simplified traffic light system from the net PN_3. Such a graph can be successfully oriented. Therefore, the decomposition process can be performed. Coloring of G_{C_3} results in three components $\mathcal{S} = \{S_1, S_2, S_3\}$:

- $S_1 = \{p_2, p_3, p_4\}$,
- $S_2 = \{p_1, p_5\}$,
- $S_3 = \{p_6\}$.

Let us now apply Algorithm 6.1 to obtain the final decomposition results. The first set S_1 remains unchanged, since achieved component is strongly connected. Therefore, S_1 forms a proper state machine component.

Two unconnected transitions are obtained during analysis of S_2. The output of t_1 and input of t_3 are not linked to any place of S_2: $O = \{t_1\}$, $I = \{t_3\}$. Thus, a nonoperational place NOP_1 is added to S_2, such that: $t_1\bullet = NOP_1 = \bullet t_3$.

Only one place p_6 is a member of S_3. Clearly, its input transition $I = \{t_5\}$ as well as output transition $O = \{t_4\}$ are unconnected. Similarly to S_2, Algorithm 6.1 supplies the net by one nonoperational place, such that $t_4\bullet = NOP_2 = \bullet t_5$.

Since all the components have been successfully supplied by NOPs, the algorithm finishes its execution. Finally, the set $\mathcal{S}$ consists of the following components:

- $S_1 = \{p_2, p_3, p_4\}$,
- $S_2 = \{p_1, p_5, NOP_1\}$,
- $S_3 = \{p_6, NOP_2\}$.

The result of decomposition is shown in Fig. 6.7. Note, that the achieved SMCs are different from the results obtained during decomposition based on the place invariants

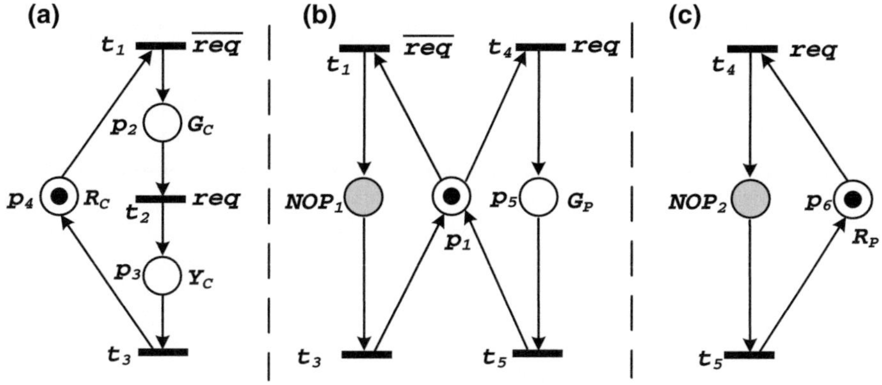

Fig. 6.7 Decomposed Petri net PN_3 (method based on the comparability graphs)

(Fig. 6.3). Let us clearly point out that both decomposed versions are correct and can be successfully implemented either as a distributed or integrated control system. The difference may influence (in case of distributed systems) the arrangement of the executable actions or (in case of integrated systems) the utilization of the devices [1, 18].

Let us briefly analyze the achieved decomposition of PN_3. Similarly to the invariant decomposition, the first component S_1 controls the lights for cars. However, S_2 is in charge of proper synchronization of the whole system (common transitions with S_1 and S_3), and additionally turns on and off green light for pedestrians. Finally, the last component S_3 controls red light for pedestrians.

Let us now decompose PN_2. We shall not present graphical illustration of the concurrency graph G_{C_2} and its TRO due to the large number of edges ($|E| = 68$). However, G_{C_2} can be transitively oriented. Furthermore, the four disjoint sets of places are obtained during coloring of G_{C_2}:

- $S_1 = \{p_1, p_4, p_5, p_{17}, p_{18}, p_{19}, p_{20}, p_{21}\}$,
- $S_2 = \{p_2, p_3, p_{14}, p_{15}, p_{16}\}$,
- $S_3 = \{p_7, p_8, p_9, p_{10}, p_{11}, p_{12}, p_{13}\}$,
- $S_4 = \{p_6\}$.

The first of the achieved components forms a strongly connected SMC, thus there is no need to apply nonoperational places to S_1. However, the analysis of S_2 results in supplementation of this sub-net by a place NOP_1, such that $t_{15}\bullet = NOP_1 = \bullet t_1$. Similarly, the third and fourth components are also complemented by one NOP: $t_{15}\bullet = NOP_2 = \bullet t_4$ (for S_3) and $t_{15}\bullet = NOP_3 = \bullet t_4$ (for S_4). Note, that all three NOPs are marked, since they substitute initially marked place p_1.

Finally, the set of decomposed components is as follows:

- $S_1 = \{p_1, p_4, p_5, p_{17}, p_{18}, p_{19}, p_{20}, p_{21}\}$,
- $S_2 = \{p_2, p_3, p_{14}, p_{15}, p_{16}, NOP_1\}$,

- $S_3 = \{p_7, p_8, p_9, p_{10}, p_{11}, p_{12}, p_{13}, NOP_2\}$,
- $S_4 = \{p_6, NOP_3\}$.

The achieved solution is very similar to the one obtained during place invariants computation and presented in Fig. 6.5. However, in the above results, the first component S_1 is in charge of setting up the main drill to the adequate position (sensor x_2) instead of controlling the movement of the wooden plank (signaled by y_1), which is done by the second component S_2. The remaining actions performed by those two components, as well as actions executed by S_3 and S_4, are exactly the same.

6.3 SM-Decomposition Based on Hypergraph Theory

A decomposition method based on the hypergraph theory includes formation of the concurrency hypergraph and further computation of exact transversals. Similarly to the idea including place invariants computation, the selection of achieved components ought to be performed.

Let us show the main idea and then illustrate it with examples.

6.3.1 *The Idea of Method*

The decomposition idea of the interpreted Petri net $PN = (P, T, F, X, Y, M_0)$ into the set of SMCs $\mathcal{S} = \{S_1, \ldots, S_n\}$ based on the hypergraph theory is divided into the following steps:

1. *Formation of the concurrency hypergraph H_C of PN.* This step is executed according to Algorithm 5.1 presented in Chap. 5.
2. *Computation of all exact transversals in H_C and formation of the set $\mathcal{S}$.* According to Theorem 5.7, an exact transversal refers to an SMC in the concurrency hypergraph. Therefore, each of achieved exact transversals forms a component in the set $\mathcal{S}$. Finally, $\mathcal{S} = \{S_1, \ldots, S_m\}$, where m is the number of all obtained SMCs.
3. *Verification if all the places of the net are covered by elements from $\mathcal{S}$.* The verification is necessary in order to check, whether PN can be decomposed with the computation of exact transversals in the concurrency hypergraph. If the place $p \in P$ is not covered by any $S \in \mathcal{S}$, the method stops execution. It means that the algorithm is unable to find the solution and different decomposition method should be used instead.
4. *Selection of state machine components in $\mathcal{S}$.* This stage is performed in the exactly same manner, as the selection of SMCs during the decomposition with the usage of place invariants (cf. Sect. 6.1.1, step 6.1.1):
   (a) *Formation of the selection hypergraph $\mathcal{H}_L = (\mathcal{S}, P)$.*

(b) *Cyclic reduction of dominated edges and dominating vertices in* $\mathcal{H}_L$.
(c) *Computation of the minimal transversal* T *in* $\mathcal{H}_L$.
(d) *Remove SMCs from* $\mathcal{S}$ *that are not indicated by* T.

5. *Replacement of repeated places by NOPs.* Similarly to the method based on p-invariants, nonoperational places are inserted in order to exchange places that exist in more than one SMC (cf. Sect. 6.1.1, step 6.1.1).

Note, that the third and fourth steps of the presented algorithm are exactly the same as in case of the method based on the place invariants computation.
Let us analyze the strong points and disadvantages of the above method.

Clearly, the main weakness of the algorithm is its computational complexity. The first two steps (formation of the reachability set, obtaining of all exact transversals) cannot be executed polynomially, since the number of reachability markings and the number of SMCs may be exponential. Thus, for some nets the final decomposition result could not be reached.

On the other hand, the obtained components are always proper and correct. In opposite to the place invariant computation, each of achieved SMCs contains exactly one initially marked place. Furthermore, the selection process (as well as replacement with NOP places) allows for choosing of the different solutions (i.e., different components among all of the achieved SMCs), which is not possible in case of the method based on the comparability graphs.

6.3.2 Examples

Let us explain the presented decomposition method with examples. Since steps 3 and 4 of the algorithm are exactly the same as for the method based on the place invariants, we shall introduce an additional real-life system for beverage production and distribution.

Initially, PN_1 will be decomposed. An incidence matrix of concurrency hypergraph $\mathcal{H}_{C_1}$ for this net is presented in Fig. 5.8 (Chap. 5). There are four exact transversals in $\mathcal{H}_{C_1}$: $X_1 = \{p_1, p_4\}$, $X_2 = \{p_3, p_6\}$, $X_3 = \{p_2, p_5\}$, $X_4 = \{p_4, p_5, p_6\}$. Each of them refers to a proper state machine component, thus the set $\mathcal{S}$ of all achieved components contains four SMCs as follows:

- $S_1 = X_1 = \{p_1, p_4\}$,
- $S_2 = X_2 = \{p_3, p_6\}$,
- $S_3 = X_3 = \{p_2, p_5\}$,
- $S_4 = X_4 = \{p_4, p_5, p_6\}$.

Note, that achieved SMCs are exactly the same, as in case of the method based on the place invariants computation (cf. Sect. 6.1.2). Therefore, the selection process leads to the final results. The initial net is decomposed into three components: $\mathcal{S} = \{S_1, S_2, S_3\}$, where

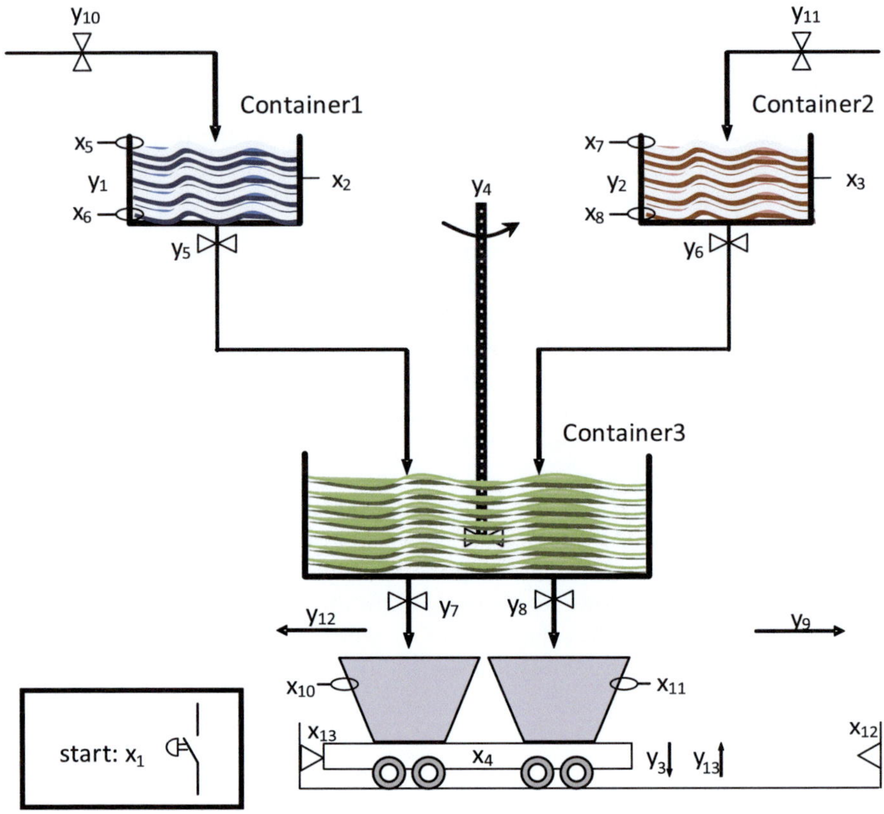

Fig. 6.8 A beverage production and distribution system

- $S_1 = \{p_1, p_4\}$,
- $S_2 = \{p_3, p_6\}$,
- $S_3 = \{p_2, p_5\}$.

Note, that graphical illustration of the decomposed components is presented in Fig. 2.8a–c.

Let us now decompose more complicated concurrent control system. We shall use a modified example taken from [19] (with minor changes shown in [9, 20]). Assume a concurrent control system for beverage production and distribution illustrated in Fig. 6.8. Let us briefly explain its functionality.

At the beginning, the system remains in an idle state, until the start button is pressed (signal x_1). It initializes the production process.

Two valves (y_{10} and y_{11}) are opened until the containers are filled up with liquid ingredients, which is noticed by sensors x_5 and y_7, respectively. If any of the containers is filled up, the ingredients are warmed up (action y_1 for *Container 1* and y_2

for *Container 2*) to achieve the adequate temperature, which is signalized by sensors x_2 and x_3.

Simultaneously to the above procedure, two cups are placed on the cart (y_3), which is signalized by sensor x_4. If both cups are already placed, the cart is moving to the left, until reaching sensor x_{13}. After that, the cart is waiting until the beverage production process is ready.

If the warm-up process is finished in both containers, their valves are opened (y_5 and y_6). The warmed components are poured to the third container, where both types of ingredients are mixed (output y_4) in order to achieve the final beverage. The liquid is prepared, until both containers are empty (sensors x_6 and x_8). Additionally, the mixer is controlled by a clock. If awaited time is elapsed, then sensor x_9 is activated.

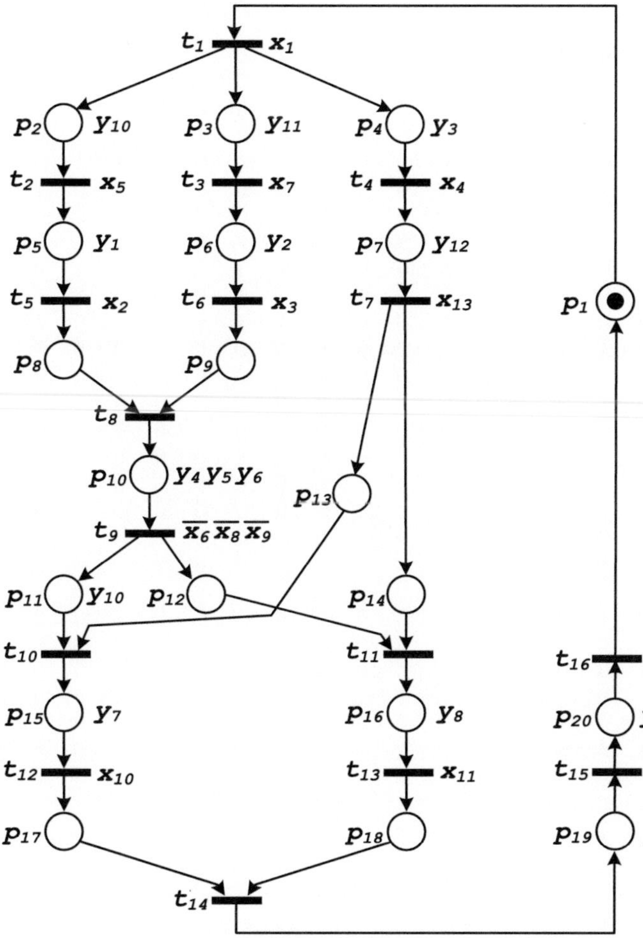

Fig. 6.9 Net PN_6 that specifies the beverage production and distribution system

p_1	p_2	p_3	p_4	p_5	p_6	p_7	p_8	p_9	p_{10}	p_{11}	p_{12}	p_{13}	p_{14}	p_{15}	p_{16}	p_{17}	p_{18}	p_{19}	p_{20}	
1	0	0	0	0	0	0	0	0	0	0	0	0	0	0	0	0	0	0	0	M_1
0	1	1	1	0	0	0	0	0	0	0	0	0	0	0	0	0	0	0	0	M_2
0	0	1	1	1	0	0	0	0	0	0	0	0	0	0	0	0	0	0	0	M_3
0	1	0	1	0	1	0	0	0	0	0	0	0	0	0	0	0	0	0	0	M_4
0	1	1	0	0	0	1	0	0	0	0	0	0	0	0	0	0	0	0	0	M_5
0	0	1	0	1	0	1	0	0	0	0	0	0	0	0	0	0	0	0	0	M_6
0	1	0	0	0	1	1	0	0	0	0	0	0	0	0	0	0	0	0	0	M_7
0	1	1	0	0	0	0	0	0	0	0	0	1	1	0	0	0	0	0	0	M_8
0	0	1	0	1	0	0	0	0	0	0	0	1	1	0	0	0	0	0	0	M_9
0	1	0	0	0	1	0	0	0	0	0	0	1	1	0	0	0	0	0	0	M_{10}
0	0	0	0	1	1	0	0	0	0	0	0	1	1	0	0	0	0	0	0	M_{11}
0	1	0	0	0	0	0	0	1	0	0	0	1	1	0	0	0	0	0	0	M_{12}
0	0	0	0	1	0	0	0	1	0	0	0	1	1	0	0	0	0	0	0	M_{13}
0	0	0	0	0	0	0	1	1	0	0	0	1	1	0	0	0	0	0	0	M_{14}
0	0	0	0	0	0	0	0	0	1	0	0	1	1	0	0	0	0	0	0	M_{15}
0	0	0	0	0	0	0	0	0	0	1	1	1	1	0	0	0	0	0	0	M_{16}
0	0	0	0	0	0	0	0	0	0	0	1	0	1	1	0	0	0	0	0	M_{17}
0	0	0	0	0	0	0	0	0	0	1	0	1	0	0	1	0	0	0	0	M_{18}
0	0	0	0	0	0	0	0	0	0	0	0	0	0	1	1	0	0	0	0	M_{19}
0	0	0	0	0	0	0	0	0	0	1	0	1	0	0	0	0	1	0	0	M_{20}
0	0	0	0	0	0	0	0	0	0	0	0	0	0	1	0	0	1	0	0	M_{21}
0	0	0	0	0	0	0	0	0	0	0	0	0	0	0	0	1	1	0	0	M_{22}
0	0	0	0	0	0	0	0	0	0	0	0	0	0	0	0	0	0	1	0	M_{23}
0	0	0	0	0	0	0	0	0	0	0	0	0	0	0	0	0	0	0	1	M_{24}
0	0	0	0	0	0	0	0	0	0	0	0	0	0	0	1	1	0	0	0	M_{25}
0	0	0	0	0	0	0	0	0	0	0	1	0	1	0	0	1	0	0	0	M_{26}
0	0	0	0	0	1	0	1	0	0	0	0	1	1	0	0	0	0	0	0	M_{27}
0	0	1	0	0	0	0	1	0	0	0	0	1	1	0	0	0	0	0	0	M_{28}
0	0	0	0	1	1	1	0	0	0	0	0	0	0	0	0	0	0	0	0	M_{29}
0	1	0	0	0	0	1	0	1	0	0	0	0	0	0	0	0	0	0	0	M_{30}
0	0	0	0	1	0	1	0	1	0	0	0	0	0	0	0	0	0	0	0	M_{31}
0	0	0	0	0	0	1	1	1	0	0	0	0	0	0	0	0	0	0	0	M_{32}
0	0	0	0	0	0	1	0	0	1	0	0	0	0	0	0	0	0	0	0	M_{33}
0	0	0	0	0	0	1	0	0	0	1	1	0	0	0	0	0	0	0	0	M_{34}
0	0	0	0	0	1	1	1	0	0	0	0	0	0	0	0	0	0	0	0	M_{35}
0	0	1	0	0	0	1	1	0	0	0	0	0	0	0	0	0	0	0	0	M_{36}
0	0	0	1	1	1	0	0	0	0	0	0	0	0	0	0	0	0	0	0	M_{37}
0	1	0	1	0	0	0	0	1	0	0	0	0	0	0	0	0	0	0	0	M_{38}
0	0	0	1	1	0	0	0	1	0	0	0	0	0	0	0	0	0	0	0	M_{39}
0	0	0	1	0	0	0	1	1	0	0	0	0	0	0	0	0	0	0	0	M_{40}
0	0	0	1	0	0	0	0	0	1	0	0	0	0	0	0	0	0	0	0	M_{41}
0	0	0	1	0	0	0	0	0	0	1	1	0	0	0	0	0	0	0	0	M_{42}
0	0	0	1	0	1	0	1	0	0	0	0	0	0	0	0	0	0	0	0	M_{43}
0	0	1	1	0	0	0	1	0	0	0	0	0	0	0	0	0	0	0	0	M_{44}

Fig. 6.10 Incidence matrix of a concurrency hypergraph for PN_5

If all three sensors (x_6, x_8 and x_9) are active, it means that the beverage is ready. Then, two valves y_7 and y_8 are opened. The liquid is being transferred to the cups located on the cart. Each cup is filled independently, until the upper limit is reached (notified by sensors x_{10} and x_{11}, respectively).

Finally, after the completion of both processes, the cart is ready for distribution of the beverage. It goes back (y_9) to the initial position, which is signaled by sensor x_{12}. Both cups are removed from the cart (y_{13}) and the system is ready for further actions awaiting for pressing of the button y_1.

The specification of the above system by an interpreted Petri net PN_6 is shown in Fig. 6.9. There are twenty places and sixteen transitions in the net. Clearly, PN_6 is an MG-net, which is live, safe, and reversible.

Let us now decompose PN_5. There are $|\mathcal{M}| = 44$ reachable markings in the net. Therefore, the concurrency hypergraph $\mathcal{H}_{C_5}$ contains 20 vertices and 44 hyperedges. Figure 6.10 presents an incidence matrix of $\mathcal{H}_{C_5}$.

At the second step of the decomposition algorithm, all exact transversals are computed. There are six exact transversals in $\mathcal{H}_{C_5}$. Each of them refers to a proper SMC

- $S_1 = \{p_1, p_2, p_5, p_8, p_{10}, p_{11}, p_{15}, p_{17}, p_{19}, p_{20}\}$,
- $S_2 = \{p_1, p_3, p_6, p_9, p_{10}, p_{11}, p_{15}, p_{17}, p_{19}, p_{20}\}$,
- $S_3 = \{p_1, p_4, p_7, p_{13}, p_{15}, p_{17}, p_{19}, p_{20}\}$,
- $S_4 = \{p_1, p_2, p_5, p_8, p_{10}, p_{12}, p_{16}, p_{18}, p_{19}, p_{20}\}$,
- $S_5 = \{p_1, p_3, p_6, p_9, p_{10}, p_{12}, p_{16}, p_{18}, p_{19}, p_{20}\}$.
- $S_6 = \{p_1, p_4, p_7, p_{14}, p_{16}, p_{18}, p_{19}, p_{20}\}$,

Clearly, all places are covered by SMCs. Let us now perform the selection process. The selection hypergraph $\mathcal{H}_{L_5} = (\mathcal{S}, P)$ consists of six vertices and twenty edges. The incidence matrix of such a hypergraph is presented in Fig. 6.11 (left). During the cyclic reduction, $\mathcal{H}_{L_5}$ is reduced to the hypergraph shown in Fig. 6.11 (right).

The first minimal transversal in $\mathcal{H}_{L_5}$ includes the following vertices: $T_1 = \{S_1, S_3, S_5, S_6\}$, Therefore, The set of SMCs after the reduction process contains four components as follows:

- $S_1 = \{p_1, p_2, p_5, p_8, p_{10}, p_{11}, p_{15}, p_{17}, p_{19}, p_{20}\}$,
- $S_3 = \{p_1, p_4, p_7, p_{13}, p_{15}, p_{17}, p_{19}, p_{20}\}$,
- $S_5 = \{p_1, p_3, p_6, p_9, p_{10}, p_{12}, p_{16}, p_{18}, p_{19}, p_{20}\}$.
- $S_6 = \{p_1, p_4, p_7, p_{14}, p_{16}, p_{18}, p_{19}, p_{20}\}$,

This time, let us adjust replacing repeated places by NOPs manually. Notice, component S_1 mainly controls a part of the system related to the left side of the system, that is *Container 1* a further filling of the left cup on the cart. Similarly, component S_5 is in charge of a proper functionality of the right side of the system (actions related to *Container 2*, filling of the right cup on the cart). Furthermore, S_6 controls the area related to the cart (its movement, releasing, and placement of cups). Let us use the above analysis in order to achieve the final decomposition (also illustrated in Fig. 6.12)

- $S_1 = \{NOP_1, p_2, p_5, p_8, p_{10}, p_{11}, p_{15}, p_{17}\}$,

$$A_{\mathcal{H}_{L_5}} = \begin{bmatrix} 1 & 1 & 1 & 1 & 1 & 1 \\ 1 & 0 & 0 & 1 & 0 & 0 \\ 0 & 1 & 0 & 0 & 1 & 0 \\ 0 & 0 & 1 & 0 & 0 & 1 \\ 1 & 0 & 0 & 1 & 0 & 0 \\ 0 & 1 & 0 & 0 & 1 & 0 \\ 0 & 0 & 1 & 0 & 0 & 1 \\ 1 & 0 & 0 & 1 & 0 & 0 \\ 0 & 1 & 0 & 0 & 1 & 0 \\ 1 & 1 & 0 & 1 & 1 & 0 \\ 1 & 1 & 0 & 0 & 0 & 0 \\ 0 & 0 & 0 & 1 & 1 & 0 \\ 0 & 0 & 1 & 0 & 0 & 0 \\ 0 & 0 & 0 & 0 & 0 & 1 \\ 1 & 1 & 1 & 0 & 0 & 0 \\ 0 & 0 & 0 & 1 & 1 & 1 \\ 1 & 1 & 1 & 0 & 0 & 0 \\ 0 & 0 & 0 & 1 & 1 & 1 \\ 1 & 1 & 1 & 1 & 1 & 1 \\ 1 & 1 & 1 & 1 & 1 & 1 \end{bmatrix} \begin{matrix} p_1 \\ p_2 \\ p_3 \\ p_4 \\ p_5 \\ p_6 \\ p_7 \\ p_8 \\ p_9 \\ p_{10} \\ p_{11} \\ p_{12} \\ p_{13} \\ p_{14} \\ p_{15} \\ p_{16} \\ p_{17} \\ p_{18} \\ p_{19} \\ p_{20} \end{matrix} \quad (\text{columns } S_1\ S_2\ S_3\ S_4\ S_5\ S_6)$$

$$A_{\mathcal{H}_{L_5}} = \begin{bmatrix} 1 & 0 & 0 & 1 & 0 & 0 \\ 0 & 1 & 0 & 0 & 1 & 0 \\ 1 & 1 & 0 & 0 & 0 & 0 \\ 0 & 0 & 0 & 1 & 1 & 0 \\ 0 & 0 & 1 & 0 & 0 & 0 \\ 0 & 0 & 0 & 0 & 0 & 1 \end{bmatrix} \begin{matrix} p_2 \\ p_3 \\ p_{11} \\ p_{12} \\ p_{13} \\ p_{14} \end{matrix} \quad (\text{columns } S_1\ S_2\ S_3\ S_4\ S_5\ S_6)$$

Fig. 6.11 Incidence matrix of $\mathcal{H}_{L_5}$ before (*left*) and after cyclic reduction (*right*)

- $S_3 = \{p_1, NOP_2, p_{13}, NOP_3\}$,
- $S_5 = \{NOP_4, p_3, p_6, p_9, NOP_5, p_{12}, p_{16}, p_{18}\}$.
- $S_6 = \{NOP_6, p_4, p_7, p_{14}, NOP_7, p_{19}, p_{20}\}$.

Notice, that places NOP_1, NOP_4, and NOP_6 are initially marked. As assumed, the first component S_1 of the decomposed net (Fig. 6.12a) controls the area of the system that is related to the first container. It manages an input valve of *Container 1* and adjusts its filling. Furthermore, the SMC also manages pouring of the beverage into the first cup on the cart. Additionally, such a process is in response of proper mixing of the ingredients.

The second component S_5 manages the second container (Fig. 6.12b). It also controls the filling of the second cup on the cart.

The third component S_6 controls movement of the cart (Fig. 6.12c). Additionally, it removes and places both cups from/on the cart.

Finally, the last component S_3 (Fig. 6.12d) is in response of the beginning of the whole process (place p_1). It also connects the first (S_1) and third (S_5) components by place p_{13}.

Note, that hypergraph $\mathcal{H}_{C_5}$ is c-exact. Therefore, achieved exact transversals $\mathcal{X} = \{X_1, \ldots, X_6\}$ form a proper sequentiality hypergraph $\mathcal{H}_{S_5}$, which is also c-exact. Moreover, further computation of all exact transversals in $\mathcal{H}_{S_5}$ results in the primary hypergraph $\mathcal{H}_{C_5}$.

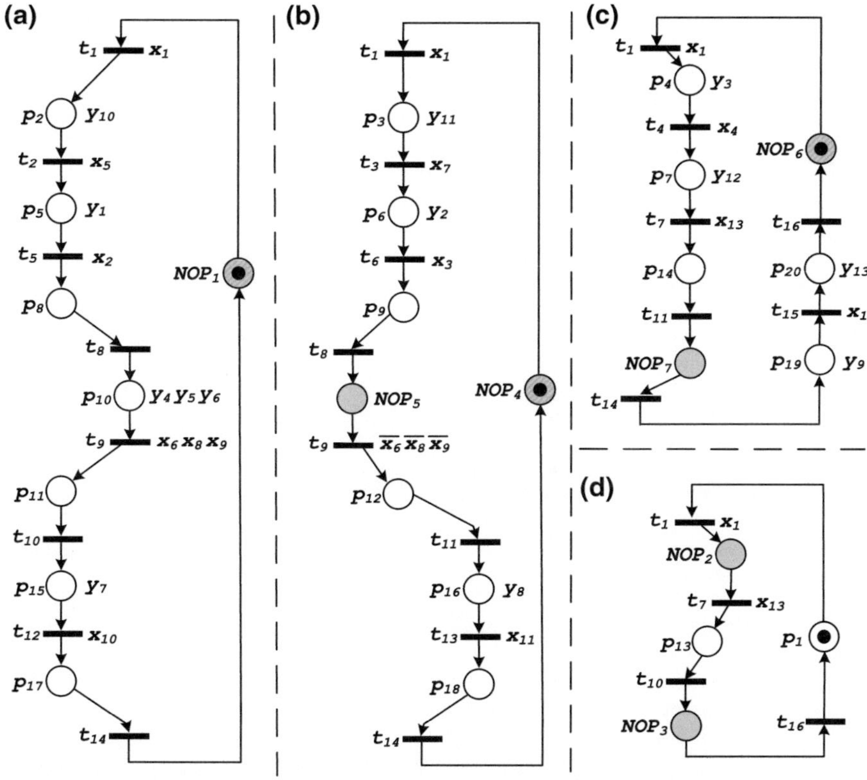

Fig. 6.12 Decomposed net PN_6 (a beverage production and distribution system)

References

1. Carmona J, Cortadella J (2006) State encoding of large asynchronous controllers. In: DAC, pp 939–944
2. Coudert O (1996) On solving covering problems. In: DAC, pp 197–202
3. Gamelin T, Greene R Introduction to topology, Dover Publications, New York
4. Garey M, Johnson D (1979) Computers and intractability: a guide to the theory of np-completeness. W.H. Freeman, New York
5. Knuth D (2000) Dancing links. In: Millennial Perspectives in Computer Science, Palgrave, pp 187–214
6. Korte B, Vygen J (2012) Combinatorial optimization: theory and algorithms, Springer New York
7. Kovalyov A (1992) Concurrency relation and the safety problem for Petri nets. In: Jensen K (ed) Proceedings of the 13th International conference on application and theory of Petri nets 1992, LNCS, vol 616. Springer, pp 299–309, June 1992
8. Pinter C (2014) A book of set theory, Dover Publications, New York
9. Roguska A (2001) Evaluation of the practical interpreted Petri nets, grafcet networks and networks based on the SEC binary control system design, 2001. Bachelors Thesis, Technical University of Zielona Góra
10. Rudell RL (1989) Logic synthesis for VLSI design. Ph.D. thesis, EECS Department, University of California, Berkeley, CA, USA

11. Sentovich E, Singh KJ, Moon CW, Savoj H, Brayton RK, Sangiovanni-Vincentelli AL (1992) Sequential circuit design using synthesis and optimization. In: ICCD '92: Proceedings of the 1991 IEEE international conference on computer design on VLSI in computer & processors, IEEE Computer Society, Washington, DC, USA, pp 328–333
12. Silva M (2013) Half a century after Carl Adam Petri's Ph.D. thesis: a perspective on the field. Annual Reviews in Control, vol 37(2). pp 191–219
13. Silva M, Terue E, Colom JM (1998) Linear algebraic and linear programming techniques for the analysis of place/transition net systems, Springer, Berlin
14. Stefanowicz Ł, Adamski M (2014) Aspects of selection of SM-components with the application of the theory of hypergraphs. In: Proceedings of the 7th IEEE international conference on human system interactions (HSI), Lisbon, Portugal, pp 221–226
15. Stefanowicz L, Adamski M, Wiśniewski R (2013) Application of an exact transversal hypergraph in selection of SM-components. In: Technological innovation for the internet of things, Springer, Heidelberg-Dordrecht, pp 250–257
16. Stefanowicz L, Adamski M, Wiśniewski R, Lipiński J (2014) Application of hypergraphs to SMCs selection. In: Technological innovation for collective awareness systems, Springer, pp 249–256
17. Stoll R (1979) Set theory and logic. Dover Publications, New York
18. Tkacz J, Adamski M (2012) Macrostate encoding of reconfigurable digital controllers from topological Petri net structure. Przeglad Elektrotechniczny 2012(8):137–140
19. Valette R (1978) Comparative study of switching representation tool with GRAFCET and Petri nets. Nouv Autom 23(12):377–382
20. Wiśniewska M (2012) Application of hypergraphs in decomposition of discrete systems, vol 23. Lecture Notes in Control and Computer ScienceUniversity of Zielona Góra Press, Zielona Góra
21. Wiśniewska M, Adamski M, Wiśniewski R (2011) Exact transversals in decomposition of Petri nets into concurrent subnets. Meas Autom Monit 58(8):851–853 in Polish
22. Wiśniewski R, Karatkevich A, Adamski M, Kur D (2014) Application of comparability graphs in decomposition of Petri nets. In: Proceedings of the 7th IEEE international conference on human system interactions (HSI), Lisbon, Portugal
23. Wiśniewski R, Stefanowicz L, Wiśniewska M, Kur D (2015) Exact cover of states in the discrete state-space system. In: 11th international conference of computational methods in sciences and engineering (ICCMSE), vol 1702 of AIP Conference Proceedings Greece, Athens, pp 1–4

Chapter 7
Prototyping of Concurrent Control Systems

7.1 Prototyping Flow of the Concurrent Systems

The prototyping flow of concurrent controllers involves several steps that lead to the final implementation as an integrated or distributed system. Such a goal can be achieved in various ways, cf. [14–17, 39, 44, 49, 58, 62, 63].

In our considerations we shall present a modified prototyping flow proposed in [30]. Such a technique joins all the important designing aspects of concurrent control systems, such as specification, decomposition (with further synchronization of achieved components), modeling, and implementation.

The general prototyping guidelines for the concurrent control systems can be divided into the following steps:

1. Specification of the concurrent control system by an interpreted Petri net.
2. Decomposition and synchronization of the concurrent control system.
3. Modeling of the decomposed modules.
4. Verification of the prototyped system.
5. Implementation of the concurrent control system.

Let us briefly describe each of the above steps.

7.1.1 Specification by an Interpreted Petri Net

Initially, the concurrent control system is specified by an interpreted Petri net (Fig. 7.1). Based on the informal descriptions, the controller is designed by the net containing places and transitions. Simultaneously marked places (states of the Petri net) permit for concurrent execution of actions. Indeed, the set of outputs (actions) of the system is associated to the net places, while the set of inputs (conditions) is tied to the transitions. Clearly, the particular action (output) is executed only if the

© Springer International Publishing Switzerland 2017

R. Wiśniewski, *Prototyping of Concurrent Control Systems Implemented in FPGA Devices*, Advances in Industrial Control,
DOI 10.1007/978-3-319-45811-3_7

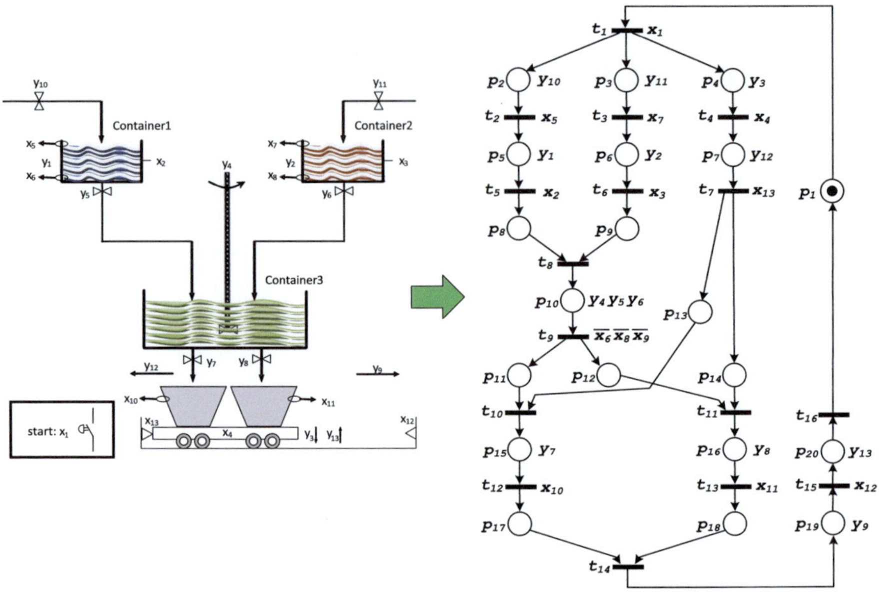

Fig. 7.1 Specification of the concurrent control system by an interpreted Petri net

place associated to it is marked. Since more than one place can be marked at the same time, there is a possibility that two or more actions are executed simultaneously.

Once the concurrent control system is specified by an interpreted Petri net, it is possible to perform formal verification of the controller [13, 22, 25, 27, 35, 59]. The specification of the system is formally verified with the use of the model checking methods [9–11, 19, 24, 26, 31, 38, 47, 51]. Such a process assures that the prototyped system meets all the user-defined requirements [28, 29].

Please note, that although the formal verification is recommended in the prototyping flow of the concurrent control systems, it is an optional step. Therefore, it shall not be considered in the further prototyping flows analyzed in the book.

7.1.2 Decomposition of the System

Decomposition permits to divide the initial system into separate modules. The most popular technique, called SM-decomposition, splits the controller into sequential automata (Fig. 7.2). Thus, each of the module can be implemented in a separate device, such as a microcontroller [4, 5], an integrated microcomputer platform [48], a programmable logic controller (PLC) [1, 40, 52], or a programmable logic device [3, 6, 37, 66]. The detailed description of the SM-decomposition can be found in Chap. 6, where three decomposition methods are proposed.

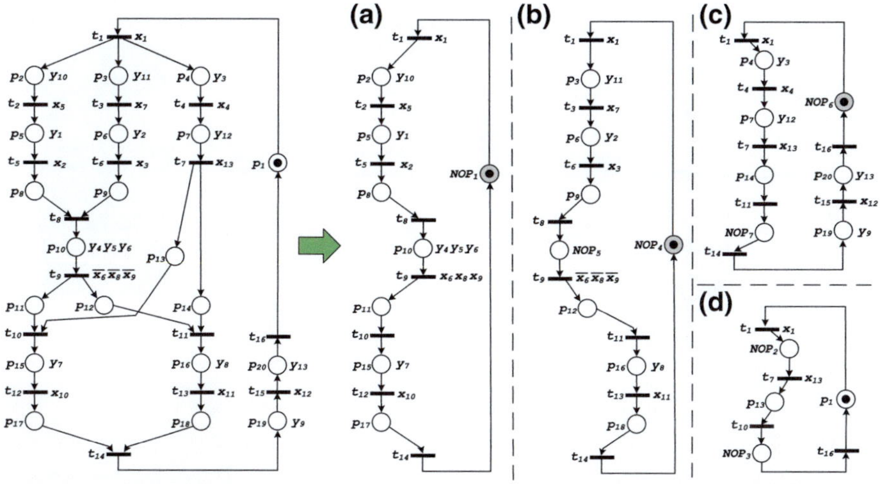

Fig. 7.2 Decomposition of the concurrent control system into sequential automata

In order to provide proper functionality of the whole system, the achieved SM components ought to be synchronized. Assume that a single transition is shared between two or more components. Clearly, before its firing it should be enabled in all the shared components.

Proper synchronization of decomposed modules is a real challenge for designers. Indeed, there are various techniques that attempt to solve such a problem [23, 32, 36, 44, 45, 60, 61].

A general synchronization concept for distributed systems is shown in [45]. The idea bases on the systems prototyped as *globally asynchronous locally synchronous (GALS)*. The decomposed modules are working in various *time domains*, that is, they are oscillated by the different clock signals.

The synchronization techniques of distributed concurrent control systems are also shown in [30]. The proposed solution applies additional places and transitions (to synchronize two components) or additional synchronization modules (for more than two components).

In our considerations (cf. Sect. 7.2) we shall focus on the integrated concurrent control systems. Moreover, all the decomposed components operate in the same time domain. It means that all SMCs are oscillated by the same clock signal. Therefore, synchronization is much simpler than in case of distributed systems and it can be easily automated (cf. Sect. 7.2.2).

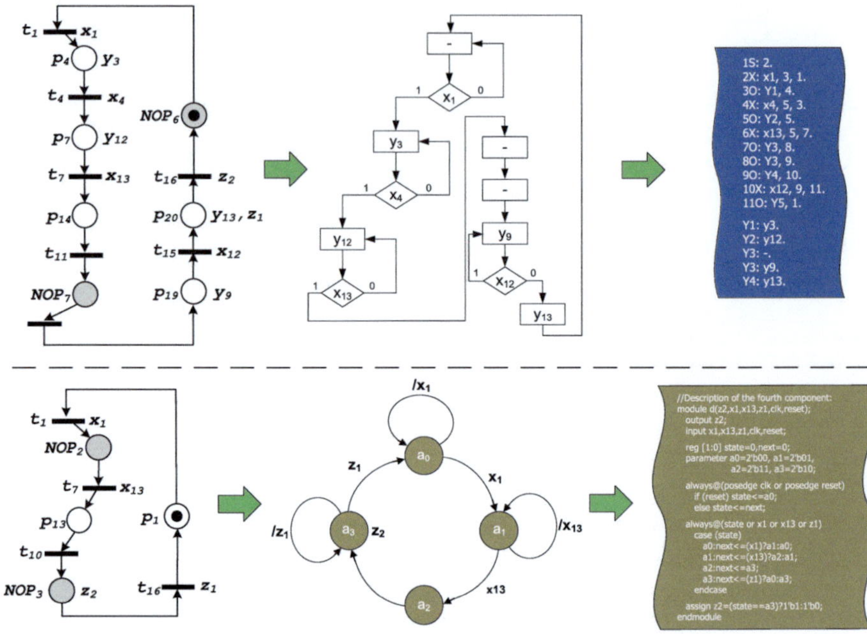

Fig. 7.3 Modeling of the decomposed and synchronized concurrent control system

7.1.3 Modeling of the Decomposed Modules

Each of the decomposed SM components can be modeled separately, with the use of different techniques. Depending on the target device, it is modeled according to the required rules (Fig. 7.3). For example, a module oriented for implementation in programmable devices is represented as a finite-state machine by a description in the hardware description languages (HDLs, cf. Chap. 8), while an SMC dedicated for a microcontroller is described by a specific programming language [30].

7.1.4 Verification of the System

Verification of the decomposed system is performed in order to achieve the properly working controller. The system is usually checked by emulation or software simulation (Fig. 7.4). Such a step can be additionally supplemented by the formal verification techniques [30].

Note that *verification* of the system checks whether the controller is properly prototyped. This step is executed after the decomposition and modeling just before the final implementation. Verification is often confused with *validation* which checks whether the adequate controller is designed:

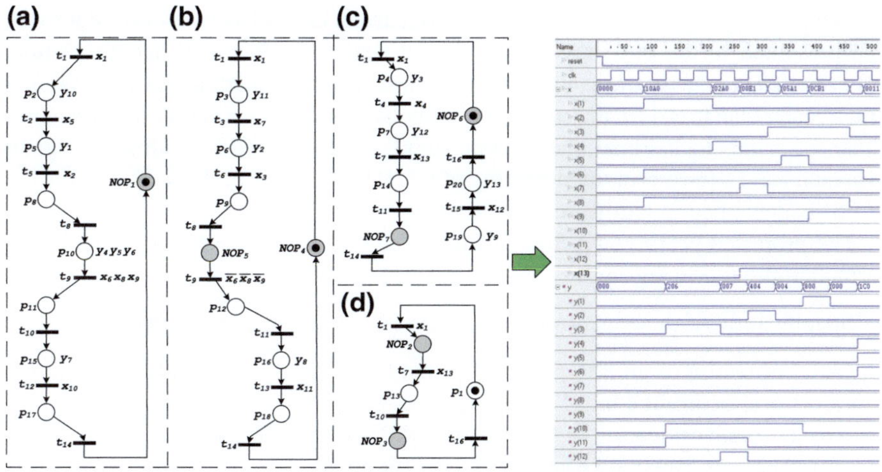

Fig. 7.4 Software simulation of the prototyped concurrent control system

- "Verification: Are we building the system right?" [21].
- "Validation: Are we building the right system?" [21].

Validation should be performed at each step of the prototyping flow. More information about verification and validation can be found in [21].

7.1.5 Implementation of the System

Finally, the concurrent system is physically implemented (programmed). Particular modules are realized with the use of various devices (distributed system) or in a single device (integrated system). Note that in case of programmable devices such as FPGAs, additional steps (logic synthesis and implementation) are ought to be performed (cf. Chap. 9).

7.2 Prototyping of Integrated Concurrent Control Systems

This section presents the prototyping guidelines for integrated concurrent control systems. It is assumed that the whole system is oscillated by the same clock signal. It is especially important due to the synchronization of the decomposed components. The prototyping method is oriented for further implementation of the system in programmable devices (mainly FPGA) with a possibility of partial reconfiguration of the controller (cf. Chap. 9).

The prototyping flow of the integrated concurrent control system is similar to the one presented in the previous section. Initial steps (specification, decomposition) are executed in exactly the same manner. However, synchronization of the system is much easier than in case of distributed systems. Since all the decomposed components share the same clock signal, they can be simply synchronized by additional signals (cf. Sect. 7.2.2). Furthermore, modeling of all components is performed with the use of FSMs described in hardware languages (cf. Sect. 7.2.3).

The proposed prototyping flow of integrated concurrent control systems is performed in the following steps:

1. Specification of the concurrent control system by an interpreted Petri net.
   (a) Formation of an interpreted Petri net based on the informal description of the concurrent control system.
   (b) Formal verification of the interpreted Petri net (optional).
   (c) Analysis of the interpreted Petri net (optional).
2. Decomposition and synchronization of the concurrent control system.
   (a) Decomposition of the concurrent control system.
   (b) Synchronization of the decomposed components.
3. Modeling of the concurrent control system.
   (a) Modeling of the decomposed modules as Moore automata (FSMs).
   (b) Description of the FSMs in hardware languages.
4. Verification of the prototyped system (software simulation).
5. Implementation (programming) of the concurrent control system.
   (a) Logic synthesis of the prototyped system.
   (b) Logic implementation of the system.
   (c) Programming data (bit-stream) generation.
   (d) Physical implementation of the system.

Let us describe the above flow illustrating with a real-life example of the simplified version of the smart home system initially shown in [30].

7.2.1 Specification by an Interpreted Petri Net

Based on the informal descriptions, the system is specified by an interpreted Petri net. This step is executed in exactly the same way, as it was shown in the previous section. Let us illustrate it with an example.

Consider a smart home system shown in Fig. 7.5 that controls the entrance of the car to the garage. Pressing of the button on the remote control by the driver (signal x_1) initializes the processing of the system. Three actions are performed simultaneously: both wings of the front gate start opening (output signals y_2 and y_4, respectively),

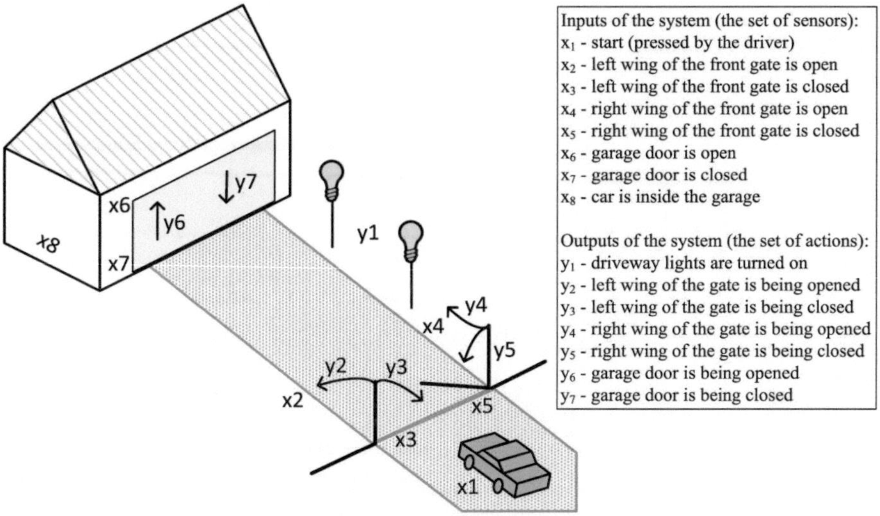

Fig. 7.5 The idea of the smart home system (entering of the car to the garage)

while the driveway lights are turned on (y_1). Each wing of the gate is being opened until the proper sensor is reached (x_2 for the left wing and x_4 for the right one).

Once the front gate is opened, the car is able to enter the home driveway. Meanwhile, the garage door is being opened (y_6). Notice that driveway lights are still turned on.

Finally, when the car enters the garage (x_8), the driveway lights are turned off ($y_1 = 0$). The wings of the front gate (y_3, y_5) and garage door are being closed until they reach the proper position (x_3, x_5 and x_7, respectively).

The above system controls only entering of the car to the garage. However, it is easy to extend the specification of the controller in order to handle the opposite direction (pulling the car out of the garage).

Figure 7.6 shows the specification of the presented system by an interpreted Petri net PN_7. The net consists of 14 places and ten transitions. It is live, safe, and reversible. The net is classified as a marked graph (MG).

Once the system is described by a Petri net, the validation ought to be performed. It should answer for the following question: "Does the achieved interpreted Petri net reflect the design assumptions"? Usually, the designer (or tester) checks whether the specified net meets all the assumed needs. For example, validation includes checking, if both wings of the front gate are open, until the car enters the driveway, etc.

Additionally, the system can be formally verified with the use of model checking technique. Note that opposite to validation such a process checks if the prototyped Petri net was properly designed [10] (analyzing, for example, reachability, liveness, safeness, etc.).

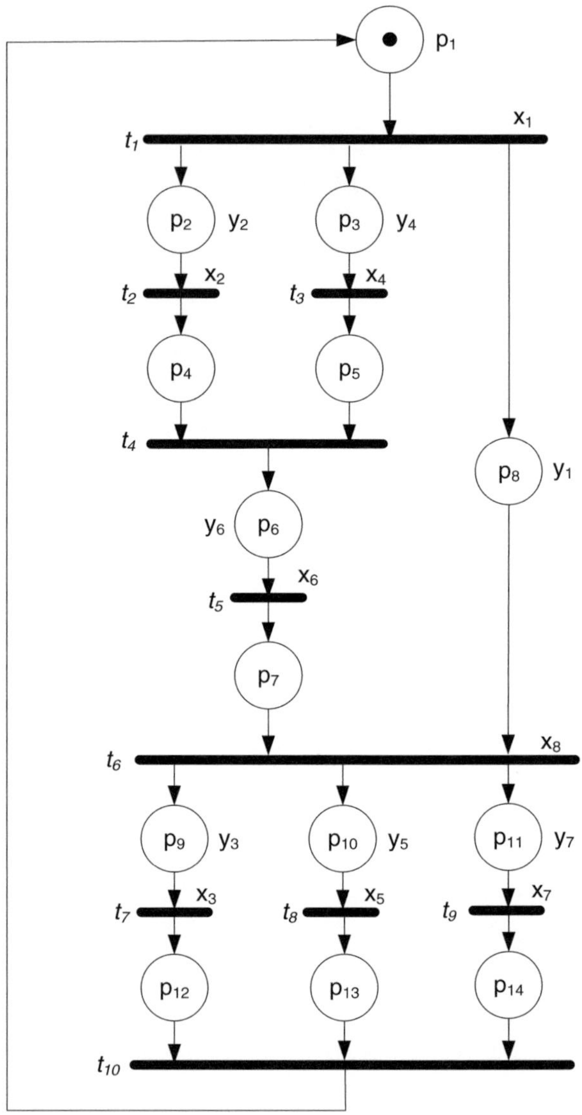

Fig. 7.6 An interpreted Petri net PN_7 of the smart home system

Optionally, the analysis of the net can be performed with the use of algorithms presented in Chap. 5. This step is especially useful in case of further partial reconfiguration of the system (cf. Chap. 9). In the presented example, analysis of PN_7 results in the following information:

- There are totally 15 reachable markings in the net.
- There are totally 9 place invariants in the net.
- There are totally 9 state machine components in the net.

- The concurrency graph of the net is a comparability graph.
- The concurrency hypergraph of the net is c-exact.
- The sequentiality hypergraph of the net is c-exact.

7.2.2 Decomposition and Synchronization of the System

The decomposition of the system is performed according to the rules shown in Chap. 6. Each of the achieved SMCs performs the sequential algorithm.

Figure 7.7 shows the decomposition results of PN_7 with the application of perfect graphs. There are three modules (components) in the decomposed net. Additionally, three NOPs were supplied during the decomposition process.

In order to ensure proper functionality of the system, the decomposed modules ought to be synchronized. As it was already mentioned, it is assumed that the whole system works in the same time domain and it is oscillated by the common clock signal (Fig. 7.8).

The synchronization concept is shown in Algorithm 7.1. The process includes analysis of all the transitions that are shared between two or more components. Additional signals $z_1, \ldots, z_i$ are used in order to synchronize shared transitions (where i is the number of all synchronization signals). Transition t in the particular component is synchronized by a logical conjunction of signals generated by input places of t in the remaining components.

Let us illustrate the synchronization idea by an example. There are four transitions to be synchronized (t_1, t_4, t_6, t_{10}) in the decomposed net PN_7. The algorithm starts with transition t_1. This transition is shared by all three SMCs, however, only $p_1 \in P$. Therefore, new output signal z_1 is assigned to p_1 and added to the set of outputs of S_1. According to the Algorithm 7.1, transition t_1 is synchronized in two remaining components (S_2 and S_3) by the logical conjunction of signals $x_1 \& z_1$, as it is shown in Fig. 7.9.

Next, transition t_4 is synchronized. It is shared by two components, S_2 and S_3. Synchronization signals z_2 and z_3 are assigned to both input places: p_5 and p_4, respectively. Furthermore, z_3 is assigned to synchronize t_4 in the component S_2, while z_2 is added to S_3.

Transitions t_6 and t_{10} are synchronized similarly. Finally, seven synchronization signals are added to the decomposed net, while input and output sets of particular SMCs are extended by the following signals (inputs and outputs of components are denoted by their number):

- S_1: $X_1 = X_1 \cup \{z_5, z_7, z_8\}$, $Y_1 = Y_1 \cup \{z_1, z_4, z_6\}$.
- S_2: $X_2 = X_2 \cup \{z_1, z_3, z_4, z_6, z_8\}$, $Y_2 = Y_2 \cup \{z_2, z_5, z_7\}$.
- S_3: $X_3 = X_3 \cup \{z_1, z_2, z_4, z_5, z_6, z_7\}$, $Y_3 = Y_3 \cup \{z_3, z_8\}$.

Figure 7.9 shows decomposed and synchronized net PN_7. At this stage, the system is additionally validated by the designer if it meets all the expected needs [21].

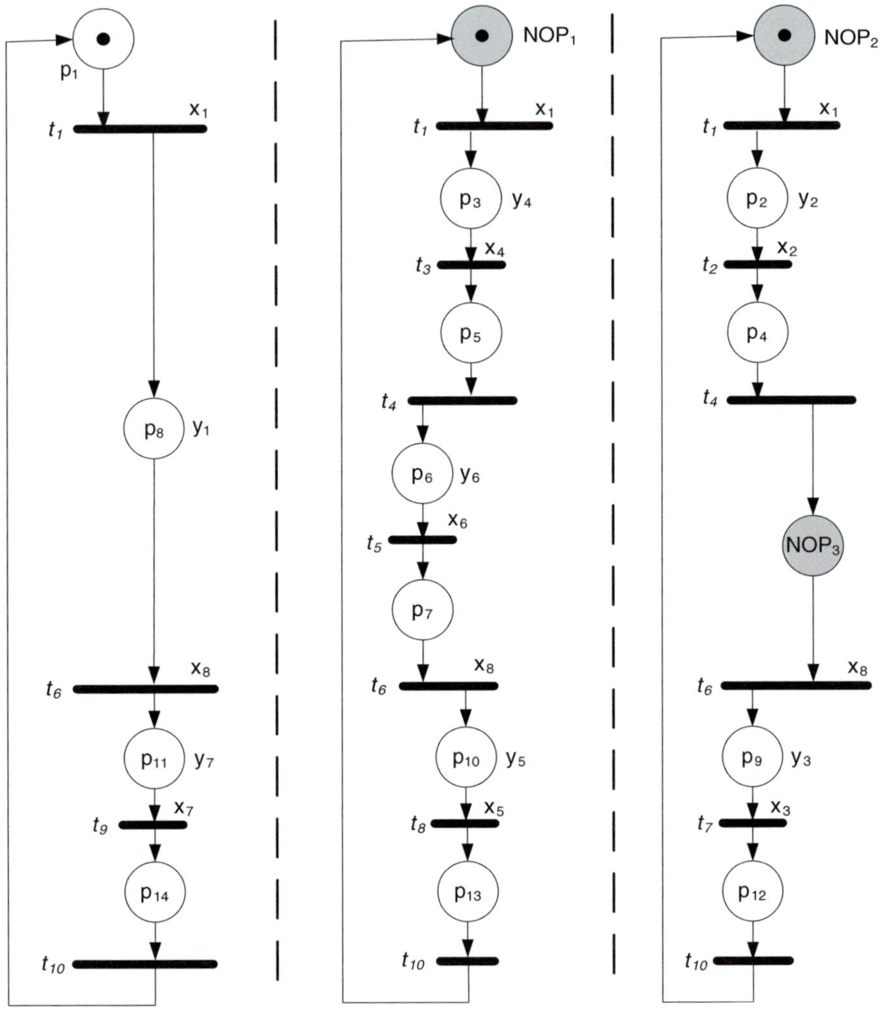

Fig. 7.7 Decomposed net PN_7

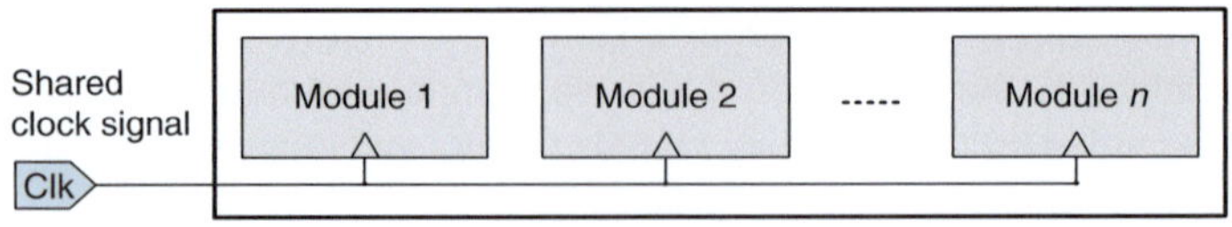

Fig. 7.8 Idea of the single time domain (decomposed modules share the clock signal)

Algorithm 7.1 Synchronization of decomposed modules (single time domain)

Input: The set of decomposed components $S \in \mathcal{S}$ of $PN = (P, T, F, M_0, X, Y)$
Output: Set $\mathcal{S}$ supplied by the synchronization signals

$i = 1$
for each $t \in T$ **do**
  if [t is shared between two or more components $S \in \mathcal{S}$] **then**
    $Z \leftarrow \emptyset$
    for each $p \in \bullet t$ such that $p \in P$ **do**
      if [there is already assigned synchronization signal z_k to p] **then**
        $Z = Z \cup \{z_k\}$
      else
        add new synchronization signal z_i and assign it to p
        $Z = Z \cup \{z_i\}$
        $i = i + 1$
      end if
    end for
    for each $S = (P', T', F', M_0, X', Y') \in \mathcal{S}$ such that [$t \in T'$] **do**
      x_t=[logical conditions assigned to $t \in T'$]
      for each $z_k \in Z$ **do**
        if [z_k is assigned to $\bullet t \in P'$] **then** $Y' = Y' \cup \{z_k\}$
        else
          $X' = X' \cup \{z_k\}$
          $x_t = x_t \,\&\, z_k$
        end if
      end for
    end for
  end if
end for

7.2.3 *Modeling of the Decomposed Modules as FSMs*

Each of the decomposed modules is modeled independently as a sequential automaton. Usually, the traditional finite state machine is used (Mealy or Moore machine [42, 43]). Additionally, advanced modeling techniques can be applied, such as *functional* or *structural* decomposition of the sequential automaton (please do not confuse with *decomposition* of concurrent systems). The first one bases on the functional description of the automaton and splits it into smaller sub-functions [18, 20, 33, 34, 41, 46, 50, 53–57]. The second method operates on the structure of the sequential controller [7, 8, 12, 65]. More details about both techniques of sequential automaton prototyping and modeling can be found in [64].

In our considerations, we shall use the traditional Moore FSM [43]. Formally, the Moore automaton is defined as a 6-tuple [64]:

$$S = (\mathcal{A}, X, Y, f, h, a_0), \tag{7.1}$$

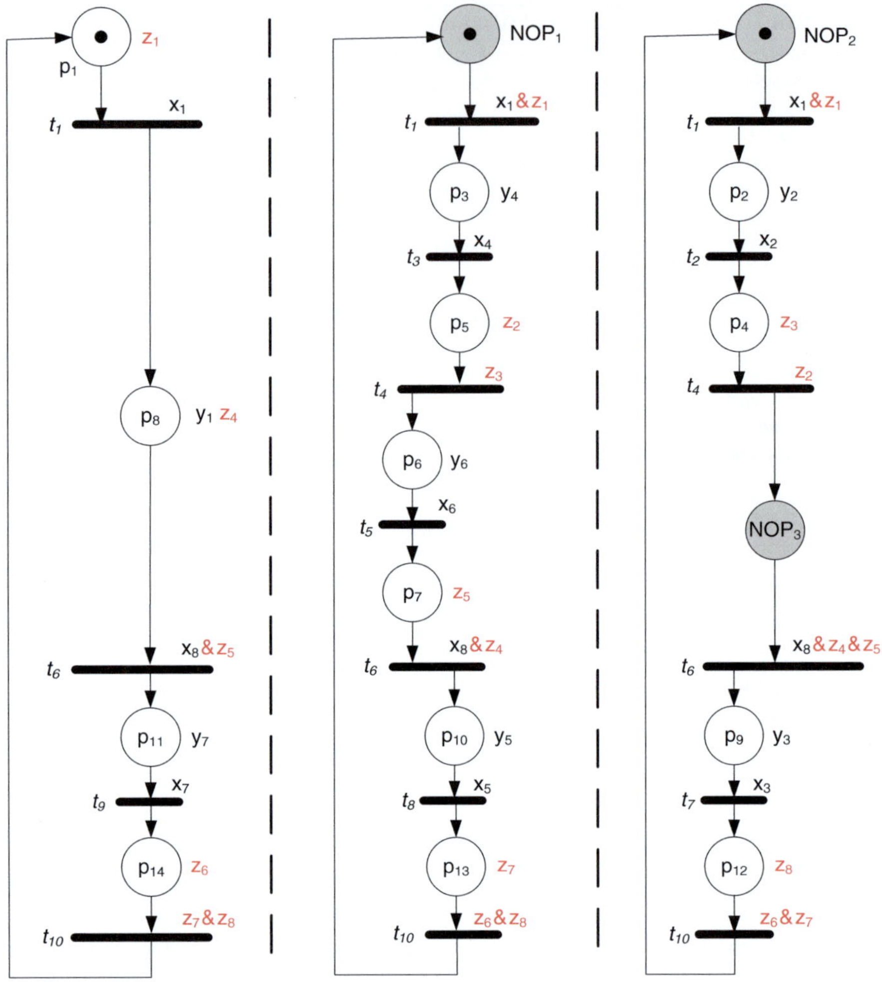

Fig. 7.9 Decomposed and synchronized net PN_7

where:

- $\mathcal{A}$ is a nonempty finite set of states;
- X is a finite set of inputs;
- Y is a finite set of outputs;
- $f : S \times X \rightarrow \mathcal{A}$ is the transition function which determines the next state $a* \in \mathcal{A}$ depending on the current state $a \in \mathcal{A}$ and on the value of the input $x \in X$;
- $h : \mathcal{A} \times X \rightarrow Y$ is the output function which determines the current output $y \in Y$, based on the current state;
- $a_0 \in \mathcal{A}$ is the initial state of an automaton.

Clearly, places of the decomposed SMC correspond to the states of the automaton. Therefore, we shall directly use place names to indicate particular state of the automaton. Furthermore, the sets of SMC inputs and outputs directly refer to the set of automaton inputs and outputs.

The automaton is usually specified formally as a *transition table* or in a graphical form, as a *state diagram*. We shall use both techniques in the example. Recall decomposed and synchronized net PN_7 that specifies the smart home system (Fig. 7.9). The initial system was decomposed into three components (modules). Let us describe each of them as a Moore automaton (Moore FSM). The set of states of achieved
automata are as follows:

- $S_1 : \mathcal{A}_1 = \{p_1, p_8, p_{11}, p_{14}\}$ (initial state: p_1).
- $S_2 : \mathcal{A}_2 = \{\mathtt{NOP}_1, p_3, p_5, p_6, p_7, p_{10}, p_{13}\}$ (initial state: $\mathtt{NOP}_1$).
- $S_3 : \mathcal{A}_3 = \{\mathtt{NOP}_2, p_2, p_4, \mathtt{NOP}_3, p_9, p_{12}\}$ (initial state: $\mathtt{NOP}_2$).

Particular modules contain the following set of inputs and outputs:

- $S_1 : X_1 = \{x_1, x_7, x_8, z_5, z_7, z_8\}, Y_1 = \{y_1, y_7, z_1, z_4, z_6\}$.
- $S_2 : X_2 = \{x_1, x_4, x_6, x_5, x_8, z_1, z_3, z_4, z_6, z_8\}, Y_2 = \{y_4, y_5, y_6, z_2, z_5, z_7\}$.
- $S_3 : X_3 = \{x_1, x_2, x_3, x_8, z_1, z_2, z_4, z_5, z_6, z_7\}, Y_3 = \{y_2, y_3, z_3, z_8\}$.

Figure 7.10 shows the state diagrams for all the decomposed components (the most left diagram refers to the component S_1, the middle to S_2, while the right one to S_3). All the components are modeled as Moore automata (Moore FSMs). Sign "&" means the logical conjunction of signals.

Equivalent transition tables for all the SMCs are shown in Table 7.1. The presented table joins three transitions tables (for component S_1, S_2 and S_3, respectively).

Note that in both the presented forms of FSMs the transitions between states are executed only if the assigned condition is fulfilled, otherwise the automaton remains in the particular state.

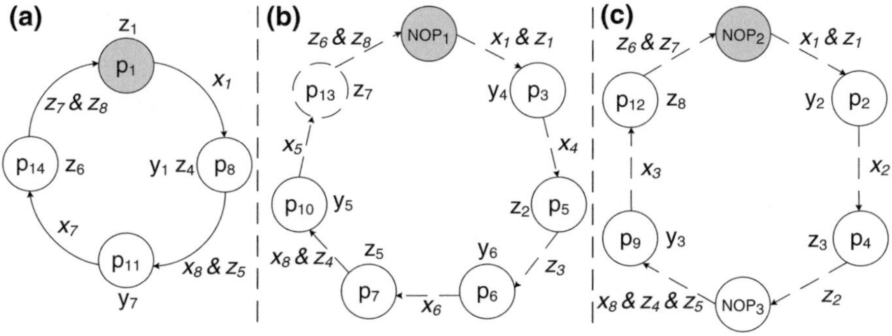

Fig. 7.10 State diagrams (Moore FSMs) of the decomposed smart home system

Table 7.1 Transition table (Moore FSMs) of the decomposed smart home system

Module	Current state	Condition	Next state	Outputs
S_1	p_1	x_1	p_8	z_1
	p_8	$x_8\&z_5$	p_{11}	y_1, z_4
	p_{11}	x_7	p_{14}	y_7
	p_{14}	$z_7\&z_8$	p_1	z_6
S_2	NOP_1	$x_1\&z_1$	p_3	–
	p_3	x_4	p_5	y_4
	p_5	z_3	p_6	z_2
	p_6	x_6	p_7	y_6
	p_7	$x_8\&z_4$	p_{10}	z_5
	p_{10}	x_5	p_{13}	y_5
	p_{13}	$z_6\&z_8$	NOP_1	z_7
S_3	NOP_2	$x_1\&z_1$	p_2	–
	p_2	x_2	p_4	y_2
	p_4	z_2	NOP_2	z_3
	NOP_3	$x_8\&z_4\&z_5$	p_9	–
	p_9	x_3	p_{12}	y_3
	p_{12}	$z_6\&z_7$	NOP_2	z_8

Similarly to the previous steps, the modeled automata are ought to be validated. Next, the prototyped FSMs are written in hardware languages. The detailed description of this step is presented in Chap. 8.

Note, that there exist tools that permit modeling FSMs in a graphical form. For example, *Active-HDL* from *Aldec* permits direct specification of the sequential automaton as a state diagram [2]. However, the model is translated to the hardware languages (Verilog or VHDL).

7.2.4 Verification of the System (Software Simulation)

Once the FSMs are described in hardware languages, they are verified by a software simulation. This step is executed similarly to the traditional flow. The particular inputs are stimulated by user-defined values, while the outputs are achieved according to the functionality of the prototyped system.

Figure 7.11 presents the results of software simulation of the smart home system performed by the tool *Active-HDL* from *Aldec*. The behavior of the controller is illustrated by signals that change values in time. Initially, *reset* is activated and the system is zeroed. When this signal is turned off, the systems starts working according to the concept shown in Fig. 7.5 and specified by an interpreted Petri net PN_7 (Fig. 7.6). Thus, after pressing of the button ($x_1 = 1$), three operations are executed simultaneously: y_1, y_2, y_4, and so on.

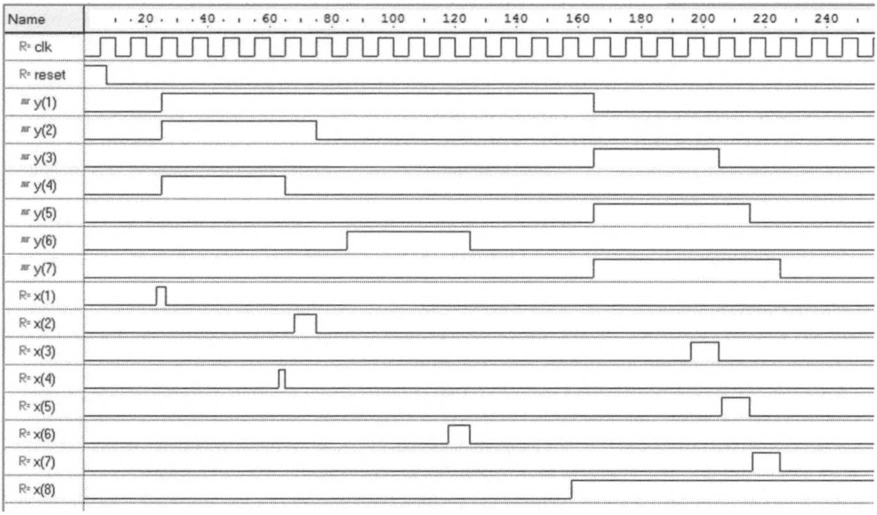

Fig. 7.11 Simulation of the prototyped smart home system

7.2.5 Implementation of the System

The final implementation of the prototyped concurrent control system consists of additional sub-steps, including logic synthesis, logic implementation, and generation of the final bit-stream, that is, the portion of data that is sent in order to program the device (for example, in case of an FPGA). Particular actions are executed strictly according to the vendor of the targeted device [3, 6, 66]. The detailed description of the implementation of the concurrent control systems in the FPGA is shown in Chap. 9.

References

1. ABB PLC homepage: http://www.abb.com/PLC. Accessed 03 April 2016
2. Aldec homepage: http://www.aldec.com. Accessed 03 April 2016
3. Altera homepage: http://www.altera.com. Accessed 03 April 2016
4. Arduino homepage: http://www.arduino.cc. Accessed 03 April 2016
5. Atmel AVR homepage: www.atmel.com/AVR. Accessed 03 April 2016
6. Atmel Programmable Logic homepage: http://www.atmel.com/products/programmable-logic. Accessed 03 April 2016
7. Adamski M, Karatkevich A, Wegrzyn M (eds) (2005) Design of embedded control systems. Springer, New York. ISBN:0-387-23630-9
8. Barkalov A, Titarenko L (2009) Logic synthesis for FSM-based control units. Lecture notes in electrical engineering, vol 53. Springer, Berlin

9. Berthomieu B, Peres F, Vernadat F (2007) Model checking bounded prioritized time Petri nets. In: ATVA'07 Proceedings of the 5th international conference on automated technology for verification and analysis
10. Best E, Esparza J (1992) Computer science logic: 5th workshop, CSL '91 Berne, Switzerland, 7–11 Oct 1991 proceedings, chapter Model checking of persistent Petri nets, pp 35–52. Springer, Berlin
11. Bérard B, Cassez F, Haddad S, Lime D, Roux O (2013) The expressive power of time Petri nets. Theoret Comput Sci 474:1–20
12. Bukowiec A (2009) Synthesis of finite state machines for FPGA devices based on architectural decomposition, vol 13. Lecture notes in control and computer science. Wydawnictwo Uniwersytetu Zielona Góra, Zielona Góra
13. Clarke E, Grumberg O, Peled D (1999) Model checking. The MIT Press
14. Cortadella J (2002) Logic synthesis for asynchronous controllers and interfaces. Springer series in Advanced microelectronics. Springer, Berlin
15. Costa A, Barbosa P, Gomes L, Ramalho F, Figueiredo J, Junior A (2010) Properties preservation in distributed execution of Petri nets models. Emerg Trends Technol Innov 314:241–250
16. Costelha H, Lima P (2007) Modelling, analysis and execution of robotic tasks using Petri nets. In: International conference on intelligent robots and systems, pp 1449–1454
17. Costelha H, Lima P (2010) Petri net robotic task plan representation: modelling, analysis and execution. In: Vedran Kordic (ed) Autonomous agents, pp 65–89. InTech
18. Czerwinski R, Kania D (2012) Area and speed oriented synthesis of FSMs for PAL-based CPLDs. Microprocess Microsyst: Embed Hardw Des 36(1):45–61
19. Darvas D, Fernandez Adiego B, Voros A, Bartha T, Blanco Vinuela E, Gonzalez Suarez V (2014) Formal verification of complex properties on PLC programs. In: Formal techniques for distributed objects, components, and systems. Lecture notes in computer science, vol 8461. Springer, Berlin, pp 284–299
20. Devadas S, Wang AR, Newton AR, Sangiovanni-Vincentelli A (1989) Boolean decomposition in multilevel logic optimization. IEEE J Solid-State Circuits 24(2):399–408
21. Easterbrook S (2016) The difference between verification and validation. http://www.easterbrook.ca/steve/2010/11/the-difference-between-verification-and-validation. Accessed 03 April 2016
22. Emerson E (2008) The beginning of model checking: a personal perspective. In: Grumberg O, Veith H (eds) 25 Years of model checking: history, achievements, perspectives. Springer, pp 27–45
23. Gomes L, Costa A, Barros JP, Lima P (2007) From Petri net models to VHDL implementation of digital controllers. In: 33rd Annual conference of the IEEE industrial electronics society, 2007, IECON 2007, pp 94–99. IEEE
24. Gourcuff V, De Smet O, Faure J-M (2006) Efficient representation for formal verification of PLC programs. In: 8th International workshop on discrete event systems, pp 182–187
25. Grobelna I (2011) Formal verification of embedded logic controller specification with computer deduction in temporal logic. Przeglad Elektrotechniczny 87(12a):47–50
26. Grobelna I (2011) Formal verification of embedded logic controller specification with computer deduction in temporal logic. Przeglad Elektrotechniczny 87(12a):40–43
27. Grobelna I (2013) Formal verification of logic controller specification by means of model checking. University of Zielona Góra Press
28. Grobelna I, Grobelny M, Adamski M (2014) Model checking of UML activity diagrams in logic controllers design. In: Proceedings of the 9th international conference on dependability and complex systems (DepCoS-RELCOMEX)
29. Grobelna I, Wiśniewska M, Wiśniewski R, Grobelny M, Mróz P (2014) Decomposition, validation and documentation of control process specification in form of a Petri net. In: Proceedings of the 7th IEEE international conference on human system interactions (HSI)

30. Grobelna I, Wiśniewski R, Grobelny M, Wiśniewska M (2016) Design and verification of real-life processes with application of Petri nets. In: IEEE Trans Syst Man Cybern: Syst. doi:10.1109/TSMC.2016.2531673
31. Hadjidj R, Boucheneb H (2009) On-the-fly tctl model checking for time Petri nets. Theoret Comput Sci 410(42):4241–4261
32. Hu H, Zhou M, Li Z (2011) Supervisor design to enforce production ratio and absence of deadlock in automated manufacturing systems. IEEE Trans Syst Man Cybern Part A: Syst Hum 41(2):201–212
33. Kania D (1999) Two-level logic synthesis on PAL-based CPLD and FPGA using decomposition. In: Proceedings. 25th EUROMICRO Conference, 1999, vol 1, pp 278–281. IEEE
34. Kania D, Kulisz J (2007) Logic synthesis for PAL-based CPLD-s based on two-stage decomposition. J Syst Softw 80(7):1129–1141
35. Kropf T (1999) Introduction to formal hardware verification: methods and tools for designing correct circuits and systems. Springer
36. Krzywicki K, Andrzejewski G (2014) Data exchange methods in distributed embedded systems. In: New trends in digital systems design, pp 126–141. VDI Verlag GmbH, Dússeldorf
37. Lattice Semiconductor homepage: http://www.latticesemi.com. Accessed 03 April 2016
38. Lampérière-Couffin S, Lesage J (2000) Formal verification of the sequential part of PLC programs. In: 5th Workshop on discrete event systems, pp 247–254
39. Leroux H, Andreu D, Godary-Dejean K (2015) Handling exceptions in Petri net-based digital architecture: from formalism to implementation on FPGAs. IEEE Trans Ind Inf 11(4):897–906
40. Mitsubishi Electric homepage: http://www.mitsubishielectric.com. Accessed 03 April 2016
41. McCluskey E (1986) Logic design principles. Prentice Hall, Englewood Cliffs, NJ
42. Mealy G (1955) A method for synthesizing sequential circuits. BSTJ 34:1045–1079
43. Moore E (1956) Gedanken experiments on sequential machines. In: Automata Studies, pp 129–153. PUP
44. Moreira M, Basilio J (2014) Bridging the gap between design and implementation of discrete-event controllers. IEEE Trans Autom Sci Eng 11(1):48–65
45. Moutinho F, Gomes L (2015) Distributed embedded controller development with Petri nets: application to globally-asynchronous locally-synchronous systems, 1st edn. Springer Publishing Company, Incorporated
46. Muthukumar V, Bignall RJ, Selvaraj H (2007) An efficient variable partitioning approach for functional decomposition of circuits. J Syst Archit 53(1):53–67
47. Penczek W, Pół rola A (2006) *Advances in verification of Time Petri Nets and timed automata.* Springer
48. Raspberry Pi homepage: www.raspberrypi.org. Accessed 03 April 2016
49. Rashid M, Anwar M, Khan A (2015) Identification of trends for model based development of embedded systems. In: 12th International symposium on programming and systems (ISPS), pp 1–8. IEEE
50. Rawski M, Jóźwiak L, Łuba T (2001) Functional decomposition with an efficient input support selection for sub-functions based on information relationship measures. J Syst Archit 47:137–155
51. Ribeiro O, Fernades J (2007) Translating synchronous Petri nets into promela for verifying behavioural properties. In: International symposium on industrial embedded systems SIES '07, pp 266–273
52. Siemens homepage: www.siemens.com. Accessed 03 April 2016
53. Sasao T (1999) Switching theory for logic synthesis. Kluwer Academic Publishers, Boston
54. Scholl C (2001) Functional decomposition with application to FPGA synthesis. Kluwer Academic Publishers, Norwell, MA
55. Sentovich E, Singh K, Lavagno L, Moon C, Murgai R, Saldanha A, Savoj H, Stephan PR, Brayton RK, Sangiovanni-Vincentelli A (1992) Sis: a system for sequential circuit synthesis. Technical report UCB/ERL M92/41, U.C. Berkeley

56. Sentovich E, Singh KJ, Moon CW, Savoj H, Brayton RK, Sangiovanni-Vincentelli AL (1992) Sequential circuit design using synthesis and optimization. In: ICCD '92: Proceedings of the 1991 IEEE international conference on computer design on VLSI in computer & processors, pp 328–333, Washington, DC, USA, 1992. IEEE Computer Society
57. Sentovich EM (1993) Sequential circuit synthesis at the gate level. PhD thesis, University of California, Berkeley, 1993. Chair-Robert K. Brayton
58. Sudacevschi V, Ababii V, Gutuleac E, Negura V (2010) HDL implementation from Petri nets description. In: 10th International conference on development and application systems, pp 236–240
59. Szpyrka M, Biernacka A, Biernacki J (2014) Methods of translation of Petri nets to nusmv language. In: International workshop on concurrency, specification and programming (CS&P), pp 245–256
60. Tanenbaum AS, Van Steen M (2007) Distributed systems. Prentice-Hall
61. Tzes A, Kim S, McShane W (1996) Applications of Petri networks to transportation network modeling. IEEE Trans Veh Technol 45(2):391–400
62. Uzam M, Burak koç I, Gelen G, Hakan Aksebzeci B (2005) Asynchronous implementation of a Petri net based discrete event control system using a Xilinx FPGA. In: Proceedings of the 35th international conference on computers and industrial engineering, pp 2025–2030
63. van der Aalst WMP (2013) Decomposing Petri nets for process mining: a generic approach. Distrib Parallel Databases 31(4):471–507
64. Wiśniewski R (2009) Synthesis of compositional microprogram control units for programmable devices, vol 14. Lecture notes in control and computer science. University of Zielona Góra Press, Zielona Góra
65. Wiśniewski R, Barkalov A, Titarenko L, Halang W (2011) Design of microprogrammed controllers to be implemented in FPGAs. Int J Appl Math Comput Sci 21(2):401–412
66. Xilinx homepage: http://www.xilinx.com. Accessed 03 April 2016

Chapter 8
Modelling of Concurrent Systems in Hardware Languages

The presented techniques include various modelling ideas presented among others in [1, 3, 6–8, 10, 11, 13, 14, 17, 21, 24]. In our considerations we shall use the Verilog language [12, 16, 18, 19, 22]. However, it can be easily transformed into VHDL (*Very High Speed Integrated Circuits Hardware Description Language*) [5, 25]. It is assumed, that Reader is familiar with the basic notations of the Verilog language.

Both modelling concepts are illustrated by the examples of real-life systems, specified by the interpreted Petri nets. Note, that the detailed descriptions of applied examples are shown in previous chapters.

It is assumed that the prototyped system is oscillated by a rising edge of an external clock signal (denoted as *clk*). Moreover, the system is zeroed by an asynchronous external *reset* signal.

8.1 Traditional Modelling of Concurrent Systems in Hardware Description Languages

The traditional modelling concept relies on the behavior of the concurrent control system [4, 20, 24]. We shall show the modelling techniques, starting from the very simple examples of an interpreted Petri net.

Initially, the main definitions are presented (for inputs, outputs, transitions, places, etc.), as well, as the description of the basic behavior of the net (transitions firing, movement of the tokens). Then, we shall move on to the more complicated interpreted Petri nets, such as description of the concurrency in the system (more than one input or/and output places of a transition) or description of conflicts (more that one input or/and output transitions of a place).

© Springer International Publishing Switzerland 2017
R. Wiśniewski, *Prototyping of Concurrent Control Systems Implemented in FPGA Devices*, Advances in Industrial Control,
DOI 10.1007/978-3-319-45811-3_8

8.1.1 The Basic Assumptions

Let us start with the basic assumptions that we use in further description of the interpreted Petri net in the Verilog language. Note, that places of the net contain tokens. Therefore, they ought to be considered as registered values, that is, declared as *reg* signals. Moreover, their changes (that is assignments to the registers) are done by the procedural assignments within the *@always* block. Furthermore, we shall oscillate all the registers by the external clock signal (active on the rising edge). The whole system is zeroed by an external and asynchronous *reset* signal.

Transitions of the described interpreted Petri net just perform simple combinational logic, i.e., *wire* signals. Thus, their modifications can be simply reached by the continuous assignments, that is *assign* statements.

Finally, outputs of the system are related to the particular places of the net. Therefore, they are described by the continuous assignments, similarly to the transitions.

8.1.2 Description of Transitions, Places, Outputs

Let us now show the description of the system specified by an interpreted Petri net in the Verilog language. Figure 8.1 shows a sample net. There are two places p_1, p_2 and two transitions t_1, t_2, while p_1 is initially marked. The presented net is live, safe, and reversible (thus it is well-formed, and can be considered as an interpreted Petri net). Clearly, the net is classified as an SM-net.

Two outputs y_1, y_2 are assigned to places p_1, p_2 (respectively), while the input signal x_1 determines firing of both transitions. Note, t_1 is enabled, when $x_1 = 1$, while t_2 is enabled, when $x_1 = 0$ (denoted as $\overline{x_1}$). The initial declaration of the above attributes in the Verilog code looks as follows:

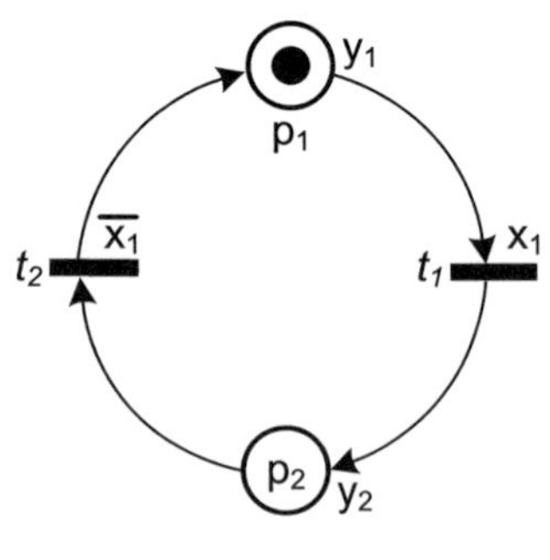

Fig. 8.1 Sample basic Petri net

```
//descriptions of all outputs and inputs:
output y1,y2;           //outputs of the net
input x1;               //input of the net
input clk, reset;       //external clock and reset

//description of the internal signals:
wire t1,t2;             //declaration of transition t1
reg p1,p2;              //declaration of places p1 and
                          p2
```

Let us now describe the behavior of the net. We shall divide the code into three main parts. The first one regards transitions (and input signals that are associated to them), the second part focuses on the places, while the last one on the outputs of the system.

There are two transitions in the net. The first one (t_1) is enabled, when p_1 is marked and the condition $x_1 = 1$ is fulfilled. Therefore, such a transition can be expressed as a logical conjunction: $t_1 = x_1 \& p_1$. Since it is a simple combinational logic, it is described by an *assign* statement. Similarly, t_2 is designed as a logical conjunction of the negation of x_1 and p_2:

```
assign t1 = x1 & p1;      //logical conjunction of
                            x1 and p1
assign t2 = ~x1 & p2;     //logical conjunction of
                            ~x1 and p2
```

Notice, that the movement of tokens between places are strictly determined by firing of transitions. Such operations can be described by a logical equations. However, places are declared as registers. Thus, procedural assignments ought to be applied. Moreover, the registers ought to be oscillated by a clock. To achieve it, we shall use an external clock source (signal *clk*). Additionally, active *reset* signal sets the net into the initial state (initial marking).

To achieve the logical equations for the places, two aspects should be considered. Clearly, a particular place becomes marked if one of its input transitions is fired. However, if the place is currently marked, the token remains untouched in such a place, unless one of its output transitions is fired. It means, that the place is marked either if it contains a token and its output transition is not fired or its input transition is fired:

```
always@(posedge clk or posedge reset)
begin
   if (reset) {p1,p2}<=2'b 10;
   else
   begin
      p1 <= p1 & ~t1 | t2;
      p2 <= p2 & ~t2 | t1;
   end
end
```

Finally, output signals are described. Note, that outputs are always tied to particular places. Therefore, they can be realized as a continuous assignment:

```
    assign y1 = p1;    //output y1 is active only if p1 is marked
    assign y2 = p2;    //output y2 is active only if p2 is marked
```

Let us combine all the above codes. The complete Verilog module description of the interpreted Petri net presented in Fig. 8.1 joins all the above partial ones, that is, descriptions of the transitions, places and outputs:

```
module net(y1,y2,x1,clk,reset);

    output y1,y2;
    input x1;
    input clk, reset;

    wire t1,t2;
    reg p1,p2;

    //equations for transitions (continuous assignments):
    assign t1 = x1 & p1;
    assign t2 = ~x1 & p2;

    always@(posedge clk or posedge reset)
    begin
        if (reset) {p1,p2}<=2'b 10;
        else
        begin
    //equations for places (procedural assignments):
            p1 <= p1 & ~t1 | t2;
            p2 <= p2 & ~t2 | t1;
        end
    end

    //equations for outputs:
    assign y1 = p1;
    assign y2 = p2;

endmodule
```

8.1.3 Description of Concurrency

Let us now describe more complicated example, presented in Fig. 8.2. The net consists of four places (with assigned five output signals to them) and three transitions that are fired according to the two input signals. Note, that in opposite to the previous example, there is a concurrency in the net. Places p_1 and p_3 are concurrent, as well as p_2 and p_4.

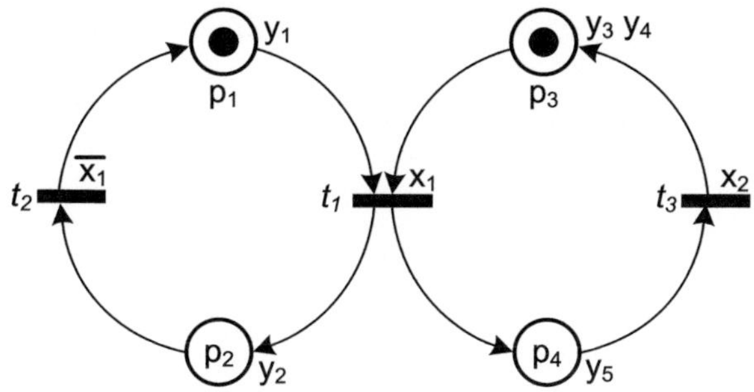

Fig. 8.2 Sample basic Petri net

Declaration of the module ports (inputs and outputs) and internal signals is similar to the previous example:

```
//descriptions of all outputs and inputs:
output y1,y2,y3,y4,y5;
input x1,x2;
input clk, reset;

//description of the internal signals:
wire t1,t2,t3;
reg p1,p2,p3,p4;
```

Let us now analyze the behavior of the transitions. The first one (t_1) is enabled when p_1 and p_3 contain tokens, and additionally, the condition $x_1 = 1$ is fulfilled. Thus, the logical conjunction of t_1 can be specified as follows: $t_1 = x_1 \& p_1 \& p_3$. Transition t_2 is enabled when p_2 is marked and if $x_1 = 0$. Finally, simultaneous marking of p_4 and the activation of signal x_2 (that is, $x_2 = 1$) enables t_3:

```
assign t1 = x1 & p1 & p3;    //conjunction of x1
                               and p1 and p3
assign t2 = ~x1 & p2;
assign t3 = x2 & p4;
```

Similarly to the previous example, the logical equations for the places are formed with the use of procedural assignments. Note, that *reset* signal returns the controller into the initial state, where places p_1 and p_3 are marked simultaneously:

```
always@(posedge clk or posedge reset)
begin
   if (reset) {p1,p2,p3,p4}<=4'b 1010;
   else
   begin
      p1 <= p1 & ~t1 | t2;
      p2 <= p2 & ~t2 | t1;
      p3 <= p3 & ~t1 | t3;
      p4 <= p4 & ~t3 | t1;
   end
end
```

The outputs are associated to the places, therefore they are simply described by continuous assignments:

```
    assign y1 = p1;
    assign y2 = p2;
    assign y3 = p3;
    assign y4 = p3;
    assign y5 = p4;
```

Eventually, the final description of the net combines all the above codes:

```
module net_concurrency(y1,y2,y3,y4,y5,x1,x2,clk,
    reset);

    output y1,y2,y3,y4,y5;
    input x1,x2;
    input clk, reset;

    wire t1,t2,t3;
    reg p1,p2,p3,p4;

    assign t1 = x1 & p1 & p3;
    assign t2 = ~x1 & p2;
    assign t3 = x2 & p4;

    always@(posedge clk or posedge reset)
    begin
        if (reset) {p1,p2,p3,p4}<=4'b 1010;
        else
        begin
            p1 <= p1 & ~t1 | t2;
            p2 <= p2 & ~t2 | t1;
            p3 <= p3 & ~t1 | t3;
            p4 <= p4 & ~t3 | t1;
        end
    end

    assign y1 = p1;
    assign y2 = p2;
    assign y3 = p3;
    assign y4 = p3;
    assign y5 = p4;

endmodule
```

8.1.4 Description of Conflicts

Conflicts is the remaining aspect of the interpreted Petri net, that has not been shown in the above descriptions. Note, that all the conflicts must be resolved at the prototyping stage, in order to classify the net as an interpreted one (cf. Chap. 2).

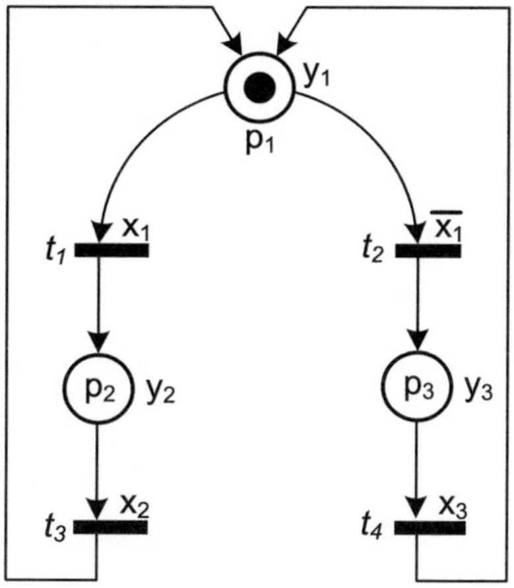

Fig. 8.3 Sample basic Petri net

Consider the interpreted Petri net shown in Fig. 8.3. There are three places and four transitions in the net. Each of three outputs is directly assigned to one of the net places, while three inputs (x_1, x_2, x_3) determine enabling and further firing of the transitions.

Let us now analyse the behavior of the net. At the initial marking, t_1 or t_2 is enabled. The enabling of those transitions is mutually exclusive, since the input signal x_1 resolves the conflict for this particular situation. Transition t_1 is enabled, when $x_1 = 1$, while t_2 is enabled when $x_1 = 0$. The remaining two transitions (t_3 and t_4) are simply formed as a logical conjunction of their input places and conditions

```
assign t1 = x1 & p1;
assign t2 = ~x1 & p1;
assign t3 = x2 & p2;
assign t4 = x3 & p3;
```

Furthermore, place p_2 becomes marked, if t_1 fires. The similar situation is with p_3. A token is moved to such a place, when t_2 is fired. However, there are two incoming arcs to p_1. It means, that such a place is marked either, when t_3 or t_4 is fired. Therefore, we shall use the logical disjunction to describe the equation for this place (and any other place that is an output of more than one transition). Additionally, when p_1 is marked, the token is kept-up in such a place unless t_1 or t_2 is enabled:

```
p1 <= p1 & ~t1 & ~t2 | t3 | t4;
p2 <= p2 & ~t3 | t1;
p3 <= p3 & ~t4 | t2;
```

Clearly, the equations for the output signals are formed in the same manner as in previous examples. The final description of the whole net is as follows:

```
module net_conflicts(y1,y2,y3,x1,x2,x3,clk,reset);

   output y1,y2,y3;
   input x1,x2,x3;
   input clk, reset;

   wire t1,t2,t3,t4;
   reg p1,p2,p3;

   assign t1 = x1 & p1;
   assign t2 = ~x1 & p1;
   assign t3 = x2 & p2;
   assign t4 = x3 & p3;

   always@(posedge clk or posedge reset)
   begin
      if (reset) {p1,p2,p3}<=3'b 100;
      else
      begin
         p1 <= p1 & ~t1 & ~t2 | t3 | t4;
         p2 <= p2 & ~t3 | t1;
         p3 <= p3 & ~t4 | t2;
      end
   end

   assign y1 = p1;
   assign y2 = p2;
   assign y3 = p3;

endmodule
```

8.1.5 Examples

The techniques presented in the previous sections allow the description of an interpreted Petri net, including concurrency and conflict resolving. Let us now show more descriptions of interpreted Petri nets in the Verilog language. We shall use the concurrent control systems from the previous chapters.

Recall the smart home system specified by the interpreted Petri net PN_7 shown in Fig. 7.6. Note, that there is a concurrency but there are no conflicts in the net. Listing 8.1 presents a Verilog code for PN_7.

Listing 8.1 Description of the home system specified by the net shown in Fig. 7.6

```
module home_system(y,x,clk,reset);

   output [1:7] y;
   input [1:8] x;
   input clk,reset;

```

```
    reg [1:14] p;
    wire [1:10] t;

    assign t[1]=x[1]&p[1];
    assign t[2]=x[2]&p[2];
    assign t[3]=x[4]&p[3];
    assign t[4]=p[4]&p[5];
    assign t[5]=x[6]&p[6];
    assign t[6]=x[8]&p[7]&p[8];
    assign t[7]=x[3]&p[9];
    assign t[8]=x[5]&p[10];
    assign t[9]=x[7]&p[11];
    assign t[10]=p[12]&p[13]&p[14];

    always@(posedge clk or posedge reset)
    begin
       if (reset) p<=14'b 10000000000000;
       else
       begin
          p[1]<=p[1]&~t[1]|t[10];
          p[2]<=p[2]&~t[2]|t[1];
          p[3]<=p[3]&~t[3]|t[1];
          p[4]<=p[4]&~t[4]|t[2];
          p[5]<=p[5]&~t[4]|t[3];
          p[6]<=p[6]&~t[5]|t[4];
          p[7]<=p[7]&~t[6]|t[5];
          p[8]<=p[8]&~t[6]|t[1];
          p[9]<=p[9]&~t[7]|t[6];
          p[10]<=p[10]&~t[8]|t[6];
          p[11]<=p[11]&~t[9]|t[6];
          p[12]<=p[12]&~t[10]|t[7];
          p[13]<=p[13]&~t[10]|t[8];
          p[14]<=p[14]&~t[10]|t[9];
       end
    end

    assign y[1]=p[8];
    assign y[2]=p[2];
    assign y[3]=p[9];
    assign y[4]=p[3];
    assign y[5]=p[10];
    assign y[6]=p[6];
    assign y[7]=p[11];
endmodule
```

Since there are 14 places and ten transitions in the net, they are grouped and declared as vectors *p* and *t*, respectively. Similarly, inputs and outputs of the net are also defined as vectors *y* and *x*.

Listing 8.2 presents the Verilog model of a simplified traffic lights system. The controller was described by the interpreted Petri net PN_3 and it was initially shown in Fig. 2.6.

Listing 8.2 Description of the simplified traffic lights in the Verilog code

```
module traffic_lights_simple(GC,YC,RC,GP,RP,req,clk,reset);

   output GC,YC,RC,GP,RP;
   input req;
   input clk,reset;

   reg [1:6] p;
   wire [1:5] t;

   assign t[1]=p[1]&p[4]&~req;
   assign t[2]=p[2]&req;
   assign t[3]=p[3];
   assign t[4]=p[1]&p[6]&req;
   assign t[5]=p[5];

   always@(posedge clk, posedge reset)
   begin
      if (reset) p<=6'b100101;
      else
      begin
         p[1]<=p[1]&~t[1]&~t[4]|t[3]|t[5];
         p[2]<=p[2]&~t[2]|t[1];
         p[3]<=p[3]&~t[3]|t[2];
         p[4]<=p[4]&~t[1]|t[3];
         p[5]<=p[5]&~t[5]|t[4];
         p[6]<=p[6]&~t[4]|t[5];
      end
   end

   assign GC=p[2];
   assign YC=p[3];
   assign RC=p[4];
   assign GP=p[5];
   assign RP=p[6];

endmodule
```

The places and transitions are declared as vectors. However, input and output signals are not grouped into vectors because of their unique names. Note, that input signal *req* resolves conflict in the net.

Finally, we shall show the description of the milling machine illustrated by the interpreted net PN_2 (cf. Fig. 2.4). Listing 8.3 presents the Verilog model of such a net.

The input and output signals are grouped into vectors, as well as internal signals representing places and transitions. Note, that assignments to the output signals are concatenated.

Listing 8.3 Description of the milling machine in Verilog

```
module milling_net(y,x,clk,reset);

   output [1:14] y;
   input [1:14] x;
   input clk,reset;

   reg [1:21] p;
   wire [1:17] t;

   assign t[1]=x[14]&p[1];
   assign t[2]=x[1]&p[2];
   assign t[3]=x[2]&p[3];
   assign t[4]=p[3]&p[5];
   assign t[5]=x[3]&p[7];
   assign t[6]=x[5]&p[9];
   assign t[7]=x[6]&p[10];
   assign t[8]=x[7]&p[11];
   assign t[9]=x[8]&p[12];
   assign t[11]=x[9]&p[14];
   assign t[12]=x[10]&p[15];
   assign t[13]=x[11]&p[17];
   assign t[14]=x[12]&p[18];
   assign t[15]=p[6]&p[13]&p[16]&p[19];
   assign t[16]=x[13]&p[20];
   assign t[17]=~x[14]&p[21];

   always@(posedge clk or posedge reset)
   begin
      if (reset) p<=14'b 10000000000000;
      else begin
         p[1]<=p[1]&~t[1]|t[17];
         p[2]<=p[2]&~t[2]|t[1];
         p[3]<=p[3]&~t[4]|t[2];
         p[4]<=p[4]&~t[3]|t[1];
         p[5]<=p[5]&~t[4]|t[3];
         p[6]<=p[6]&~t[15]|t[4];
         p[7]<=p[7]&~t[5]|t[4];
         p[8]<=p[8]&~t[6]|t[5];
         p[9]<=p[9]&~t[7]|t[6];
         p[10]<=p[10]&~t[8]|t[7];
         p[11]<=p[11]&~t[9]|t[8];
         p[12]<=p[12]&~t[10]|t[9];
         p[13]<=p[13]&~t[15]|t[10];
         p[14]<=p[14]&~t[11]|t[4];
         p[15]<=p[15]&~t[12]|t[11];
         p[16]<=p[16]&~t[15]|t[12];
         p[17]<=p[17]&~t[13]|t[4];
         p[18]<=p[18]&~t[14]|t[13];
         p[19]<=p[19]&~t[15]|t[14];
         p[20]<=p[20]&~t[16]|t[15];
         p[21]<=p[21]&~t[17]|t[16];
      end
```

```
    end

    assign y={p[2],p[4],p[6:12],p[14],p[15],p[17],p
        [18],p[20]};

endmodule
```

8.2 Modelling of Concurrent Systems as a Composition of Sequential Automata

This section describes the modelling idea of concurrent control systems seen as a composition of sequential automata. It is assumed that the initial controller has been already split into sequential modules with the use of decomposition methods shown in Chap. 6. Furthermore, the proper synchronization of the achieved components is assured, according to the technique proposed in the previous chapter (cf. Sect. 7.2.3).

We shall begin with the basic description of an FSM in the Verilog language. Next, a few examples will be shown. All of them based on the real-life concurrent control systems presented in the previous chapters.

8.2.1 Description of an FSM in Verilog

Let us start with the basic assumptions regarding description of a Moore FSM in Verilog. Figure 8.4 shows the sample specification of a synchronous reversible counter. Two outputs y_2 and y_1 indicate the current (binary) value of the counter, while input x specifies the counting direction: forward ($x = 1$) or reverse ($x = 0$).

Clearly, the counter requires at least two registers for encoding of four internal states. In our considerations we shall use the *Gray* code. Furthermore, *D flip-flops* with the asynchronous reset are used (taken directly from the Xilinx primitives and denoted as *FDC* [23]). The typical prototyping method of the Moore FSM [2, 9, 15, 22] leads to the following equations of the excitation functions for flip-flops:

- $d_1 = q_2 \oplus x$,
- $d_2 = q_1 \otimes x$,

where d_1, d_2 denote registers inputs, q_1, q_2 refer to their outputs, sign $\oplus$ means logic *exclusive-or* (*xor*), while $\otimes$ denotes its logical complement (*xnor*).

Furthermore, the equations for the outputs are as follows:

- $y_1 = q_2 \oplus q_1$,
- $y_2 = q_2$.

Fig. 8.4 Sample basic Petri net

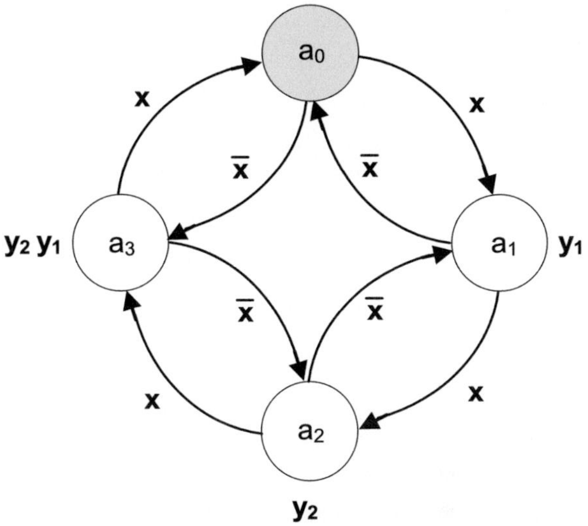

The final structural diagram of the counter is shown in Fig. 8.5. Let us describe such a model with the use of Verilog language. Listing 8.4 shows the sample code of the counter realized in a structural way.

Listing 8.4 Description of the counter in Verilog (structural)

```
module counter_structural(y1,y2,x,clk,reset);

     output y1,y2;
     input x;
     input clk, reset;

     wire d1,d2;
     wire q1,q2;

     assign d1=x^q1;
     assign d2=x^~q2;

     FDC fd1 (q1,clk,reset,d1);
     FDC fd2 (q2,clk,reset,d2);

     assign y1=q1^q2;
     assign y2=q2;

endmodule
```

Both flip-flops are described as an instantiation of the primitive *FDC* from the *Xilinx* library. Equations for register inputs, as well as equations for the outputs are realized by the continuous assignments.

The above code presents the structural description of the FSM in Verilog. Alternatively, an automaton can be realized with the use of behavioral statements. Listing 8.5

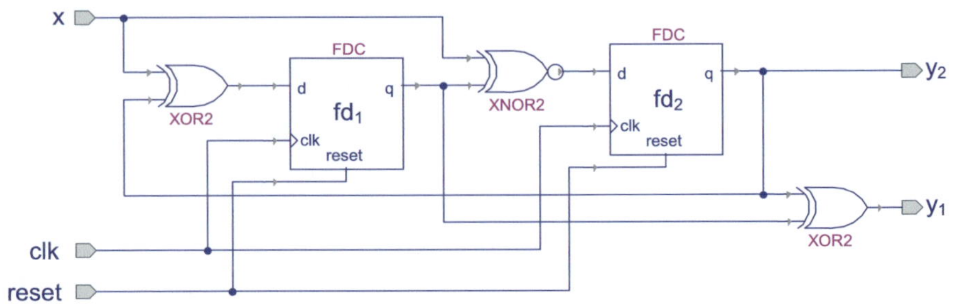

Fig. 8.5 Sample basic Petri net

presents an exemplary behavioral description of the counter in the Verilog code. Note, that such a description can be performed in a different way. In our considerations we use *case* statement.

Listing 8.5 Description of the counter in Verilog (behavioral)

```
module counter (y1,y2,x,clk,reset);

    output y1,y2;
    input x;
    input clk, reset;

    reg [1:0] state,next; //declaration of current and next states
    parameter a0=2'b00, a1=2'b01, a2=2'b11, a3=2'b10; //Gray code

    always @(posedge clk or posedge reset)
    if (reset) state <= a0;
    else state <= next;

    always @(state or x) //computation of the next state
    case(state)
        a0: next <= (x)?a1:a3;
        a1: next <= (x)?a2:a0;
        a2: next <= (x)?a3:a1;
        a3: next <= (x)?a0:a2;
        default: next <= a0;
    endcase

    assign y1=(state==a1||state==a3)?1'b1:1'b0;
    assign y2=(state==a2||state==a3)?1'b1:1'b0;

endmodule
```

The behavioral description of an FSM includes three main blocks, two of them are procedural, while the remaining one is continuous. The first procedural block is active on the rising edges of the clock and reset signals. It zeroes the system (if *reset* is active) or switches the current state of the automaton into the next one. The second *always* block performs assignments of the next state, based on the input signal and current state. Finally, the continuous assignments perform the equations for outputs.

Note, that there is a possibility for specific encoding of the states. In the above example, particular states of the automaton are encoded with the use of *Gray* code.

Note, that both presented FSM descriptions (structural and behavioral) are fully synthesizable and can be successfully used in the prototyping flow of the concurrent control systems presented in this book. However, in the further examples we shall use the behavioral description of FSMs.

8.2.2 *Examples*

Let us now present examples of concurrent control systems described in Verilog as a composition of sequential automata. At the beginning, the smart home system shall be presented. According to Fig. 7.10 (or Table 7.1) there are three FSMs in the decomposed system. Let us apply the technique presented in the previous section to describe the first state machine component in the Verilog language.

Listing 8.6 Description of the first component of the smart home system

```
//Description of the first component:
module S1 (y1,y7,z1,z4,z6,x1,x7,x8,z5,z7,z8,clk,reset);

   output y1,y7,z1,z4,z6;
   input x1,x7,x8,z5,z7,z8,clk,reset;

   reg [1:0] state=0,next=0;
   parameter p1=2'b00, p8=2'b01, p11=2'b11, p14=2'b10;

   always@(posedge clk or posedge reset)
   if (reset) state <= p1;
   else state <= next;

   always@(state or x1 or x7 or x8 or z5 or z7 or z8)
   case (state)
      p1:next <= (x1)?p8:p1;
      p8:next <= (x8&&z5)?p11:p8;
      p11:next <= (x7)?p14:p11;
      p14:next <= (z7&&z8)?p1:p14;
      default:next<=p1;
   endcase

   assign z1=(state==p1)?1'b1:1'b0;
   assign y1=(state==p8)?1'b1:1'b0;
   assign z4=(state==p8)?1'b1:1'b0;
```

```
    assign y7=(state==p11)?1'b1:1'b0;
    assign z6=(state==p14)?1'b1:1'b0;

endmodule
```

Listing 8.6 shows the sample code of S_1. Note, that there are six additional synchronization signals (three outputs: z_1, z_4, z_6 and three inputs: z_5, z_7, z_8). Such signals are used for proper communication with other two components (cf. Sect. 7.2.2).

Internal states of the automaton are encoded with the use of *Gray* code. Their labels (names) refer directly to the places in the decomposed Petri net. Therefore, the initial state is denoted as p_1 (a place that holds a token in the initial marking).

Listing 8.7 presents the Verilog description of two remaining components. They are modelled in the same manner as the first SMC. Note, that in both components three registers are applied to keep the value of automaton states.

Listing 8.7 Description of the remaining components of the home system

```
//Description of the second component:
module S2(y4,y5,y6,z2,z5,z7,x1,x4,x5,x6,x8,z1,z3,z4,z6,
    z8,clk,reset);

    output y4,y5,y6,z2,z5,z7;
    input x1,x4,x5,x6,x8,z1,z3,z4,z6,z8,clk,reset;

    reg [2:0] state=0,next=0;
    parameter NOP1=3'b000, p3=3'b001, p5=3'b011, p6=3'
        b010, p7=3'b110, p10=3'b111, p13=3'b101;

    always@(posedge clk or posedge reset)
    if (reset) state <= NOP1; else state <= next;

    always@(state or x1 or x4 or x5 or x6 or x8
            or z1 or z3 or z4 or z6 or z8)
    case (state)
       NOP1:next<=(x1&z1)?p3:NOP1;
       p3:next<=(x4)?p5:p3;
       p5:next<=(z3)?p6:p5;
       p6:next<=(x6)?p7:p6;
       p7:next<=(x8&&z4)?p10:p7;
       p10:next<=(x5)?p13:p10;
       p13:next<=(z6&&z8)?NOP1:p13;
       default:next<=NOP1;
    endcase

    assign y4=(state==p3)?1'b1:1'b0;
    assign z2=(state==p5)?1'b1:1'b0;
    assign y6=(state==p6)?1'b1:1'b0;
    assign z5=(state==p7)?1'b1:1'b0;
    assign y5=(state==p10)?1'b1:1'b0;
    assign z7=(state==p13)?1'b1:1'b0;

endmodule

//Description of the third component:
```

```
module S3 (y2,y3,z3,z8,x1,x2,x3,x8,z1,z2,z4,z5,z6,z7,
    clk,reset);

   output y2,y3,z3,z8;
   input x1,x2,x3,x8,z1,z2,z4,z5,z6,z7,clk,reset;

   reg [2:0] state=0,next=0;
   parameter NOP2=3'b000, p2=3'b001, p4=3'b011, NOP3=3'
      b010, p9=3'b110, p12=3'b100;

   always@(posedge clk or posedge reset)
   if (reset) state <= NOP2; else state <= next;

   always@(state or x1 or x2 or x3 or x8 or z1
           or z2 or z4 or z5 or z6 or z7)
   case (state)
      NOP2:next<=(x1&z1)?p2:NOP2;
      p2:next<=(x2)?p4:p2;
      p4:next<=(z2)?NOP3:p4;
      NOP3:next<=(x8&z4&z5)?p9:NOP3;
      p9:next<=(x3)?p12:p9;
      p12:next<=(z6&&z7)?NOP2:p12;
      default:next<=NOP2;
   endcase

   assign y2=(state==p2)?1'b1:1'b0;
   assign z3=(state==p4)?1'b1:1'b0;
   assign y3=(state==p9)?1'b1:1'b0;
   assign z8=(state==p12)?1'b1:1'b0;

endmodule
```

Finally, the main module of the decomposed system ought to be prepared. Such a module (often called *top module*) contains instantiations of all three components. Listing 8.8 presents the description of the main module for the decomposed smart home system.

Listing 8.8 Description of the main (top) module of the home system

```
//Main module:
module home_system_decomposed (y,x,clk,reset);

   output [1:7] y;
   input [1:8] x;
   input clk,reset;

   wire z[1:8];

   S1 SMC1 (y[1],y[7],z[1],z[4],z[6],x[1],x[7],x[8],z
      [5],z[7], z[8],clk,reset);
   S2 SMC2 (y[4],y[5],y[6],z[2],z[5],z[7],x[1],x[4],x
      [5],x[6], x[8],z[1],z[3],z[4],z[6],z[8],clk,
      reset);
   S3 SMC3 (y[2],y[3],z[3],z[8],x[1],x[2],x[3],x[8],z
      [1],z[2], z[4],z[5],z[6],z[7],clk,reset);

endmodule
```

Let us now present the description of the simplified traffic lights system. Recall the results of decomposition based on the comparability graphs (cf. Fig. 6.7). In order to describe the decomposed system in Verilog, proper synchronization ought to be assured. Figure 8.6 shows decomposed and synchronized net PN_3.

There are three components in the decomposed net and five synchronization signals. Note, that the second module (Fig. 8.6) contains a conflict which is resolved by the input signal *req*. Listing 8.9 presents a sample Verilog code of all the components of the decomposed traffic light system.

Listing 8.9 Description of the decomposed components of the traffic light system

```
module S1 (GC,YC,RC,z1,z3,req,z2,clk,reset);

    output GC,YC,RC,z1,z3;
    input req,z2,clk,reset;

    reg [1:0] state=0,next=0;
    parameter p4=2'b00, p2=2'b01, p3=2'b11;

    always@(posedge clk or posedge reset)
    if (reset) state<=p4; else state<=next;

    always@(state or req or z2)
    case (state)
        p4:next<=(!req&&z2)?p2:p4;
        p2:next<=(req)?p3:p2;
        p3:next<=p4;
        default:next<=p4;
    endcase

    assign {RC,z1}=(state==p4)?2'b11:2'b00;
    assign GC=(state==p2)?1'b1:1'b0;
    assign {YC,z3}=(state==p3)?2'b11:2'b00;

endmodule

module S2 (GP,z2,z5,req,z1,z3,z4,clk,reset);

    output GP,z2,z5;
    input req,z1,z3,z4,clk,reset;

    reg [1:0] state=0,next=0;

    parameter p1=2'b00, NOP1=2'b01, p5=2'b10;

    always@(posedge clk or posedge reset)
    if (reset) state<=p1; else state<=next;

    always@(state or req or z1 or z3 or z4)
    case (state)
        p1:next<=(!req&&z1)?NOP1:(req&&z4)?p5:p1;
        NOP1:next<=(z3)?p1:NOP1;
```

```
        p5:next<=p1;
        default:next<=p1;
    endcase

    assign z2=(state==p1)?1'b1:1'b0;
    assign {GP,z5}=(state==p5)?2'b11:2'b00;

endmodule

module S3 (RP,z4,req,z2,z5,clk,reset);

    output RP,z4;
    input req,z2,z5,clk,reset;

    reg state=0,next=0; parameter p6=1'b1, NOP2=1'b0;

    always@(posedge clk or posedge reset)
    if (reset) state<=p6; else state<=next;

    always@(state or req or z2 or z5)
    case (state)
        p6:next<=(req&&z2)?NOP2:p6;
        NOP2:next<=(z5)?p6:NOP2;
        default:next<=p6;
    endcase

    assign {RP,z4}=(state==p6)?2'b11:2'b00;

endmodule
```

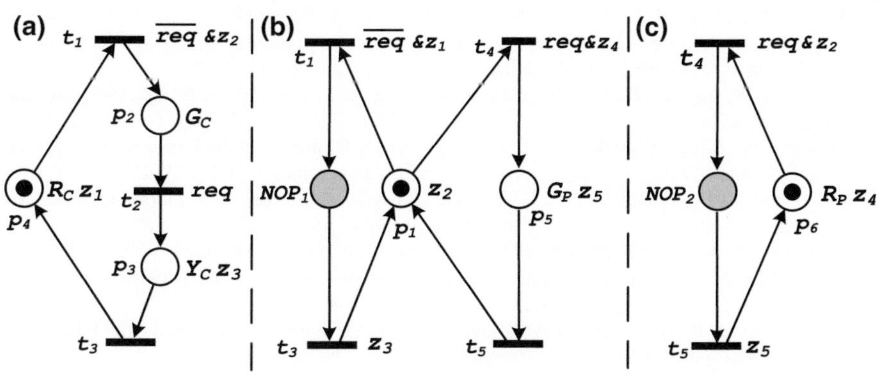

Fig. 8.6 Synchronized net PN_3 after the decomposition (comparability graphs)

Finally, the main module joins all three components. Listing 8.10 illustrates the Verilog code of the top module for the decomposed traffic light system.

Listing 8.10 Description of the main module of the decomposed traffic light system

```
module traffic_lights_simple_dec (GC,YC,RC,GP,RP,req
    ,clk,reset);

    output GC,YC,RC,GP,RP;
    input req;
    input clk,reset;

    wire z[1:5];

    S1 SMC1 (GC,YC,RC,z[1],z[3],req,z[2],clk,reset);
    S2 SMC2 (GP,z[2],z[5],req,z[1],z[3],z[4],clk,
             reset);
    S3 SMC3 (RP,z[4],req,z[2],z[5],clk,reset);

endmodule
```

References

1. Adamski M, Wegrzyn M (2009) Petri nets mapping into reconfigurable logic controllers. Electron Telecommun Q 55(2):157–182
2. Barkalov A, Titarenko L (2009) Logic synthesis for FSM-based control units, vol 53. Lecture Notes in Electrical Engineering, Springer, Berlin
3. Bazydlo G, Adamski M (2011) Specification of UML 2.4 HSM and its computer based implementation by means of Verilog. Przeglad Elektrotechniczny 87(11):145–149 in Polish
4. Blunno I, Lavagno L (2000) Automated synthesis of micro-pipelines from behavioral verilog HDL. In: (ASYNC 2000) Proceedings of the sixth international symposium on advanced research in asynchronous circuits and systems, 2000, IEEE, pp 84–92
5. Brown S, Vernesic Z (2000) Fundamentals of digital logic with VHDL design. McGraw Hill, New York, USA
6. Bukowiec A, Tkacz J, Gratkowski T, Gidlewicz T (2013) Implementation of algorithm of Petri nets distributed synthesis into FPGA. Int J Electron Telecommun 59:317–324
7. Bukowiec A, Doligalski M (2013) Petri net dynamic partial reconfiguration in FPGA. In: Computer aided systems theory-EUROCAST, Springer, pp 436–443
8. Bukowiec A, Mróz P (2012) An FPGA synthesis of the distributed control systems designed with Petri nets. In: Proceedings of the IEEE 3rd international conference on networked embedded systems for every application, Liverpool, UK, pp 1–6
9. Czerwinski R, Kania D (2012) Area and speed oriented synthesis of FSMs for PAL-based CPLDs. Microprocess Microsyst—Embed Hardw Des 36(1):45–61
10. Gomes L, Costa A, Barros JP, Lima P (2007) From Petri net models to VHDL implementation of digital controllers. In: IECON 2007. 33rd annual conference of the IEEE industrial electronics society, 2007, IEEE, pp 94–99
11. Grobelna I (2013) Formal verification of logic controller specification by means of model checking. University of Zielona Góra Press
12. IEEE Standard for Verilog Hardware Description Language (2006)
13. Mallet F, Gaffé D, Boéri F (2000) Concurrent control systems: from grafcet to VHDL. In: EUROMICRO, IEEE Computer Society, pp 1230–1234

14. Minns P, Elliott I (2008) FSM-based digital design using Verilog HDL, Wiley
15. Moore E (1956) Gedanken experiments on sequential machines. In: Automata studies, PUP, pp 129–153
16. Palnitkar S (2003) Verilog HDL: a guide to digital design and synthesis, vol 1. Prentice Hall Professional
17. Sudacevschi V, Ababii V, Gutuleac E, Negura V (2010) HDL implementation from Petri nets description. In: 10th International conference on development and application systems, pp 236–240
18. Thomas D, Moorby P (2002) The verilog hardware description language, 5th edn. Kluwer Academic Publishers, Norwell, MA
19. Thomas D, Moorby P (2008) The Verilog hardware description language, Springer Science & Business Media
20. Vanbekbergen P, Wang A, Keutzer K (1995) A design and validation system for asynchronous circuits. In: Proceedings of the 32nd annual ACM/IEEE design automation conference, ACM, pp 725–730
21. Wegrzyn M (2006) Petri net decomposition approach for partial reconfiguration of logic controllers. In: Proceedings of the workshop on discrete-event system design, Rydzyna, Poland, pp 323–328
22. Wiśniewski R (2009) Synthesis of compositional microprogram control units for programmable devices, vol 14. Lecture Notes in Control and Computer ScienceUniversity of Zielona Góra Press, Zielona Góra
23. Xilinx primitives (Virtex-5 family). http://www.xilinx.com/support/documentation/sw_manuals/xilinx11/virtex5_scm.pdf. Accessed 04 March 2016
24. Yakovlev A, Gomes L, Lavagno L (2000) Hardware design and Petri nets, Springer
25. Zwolinski M (2000) Digital system design with VHDL. Addison-Wesley Longman Publishing Co. Inc, Boston, MA, USA

Chapter 9
Implementation of Concurrent Control Systems in FPGA

9.1 Introduction to the Programmable Devices

The idea of digital circuits and programmable devices dates back to the late 1940s, when the first transistor was prototyped as a point-contact device, initially formed from germanium. Such an invention was essential for further logic devices. The next decade benefitted from the development of the first digital gates and circuits–so-called *transistor–transistor logic* (*TTL*) device. Such devices consist of up to sixteen input/output pins. Each of them performs a simple digital function. For example the device *7400* contained four 2-input NAND gates, *7404*—six NOT inverters, and so on [22]. Those circuits were the first devices called *Application Specific Integrated Circuit (ASIC)*. Logic functions of such a device were fixed and unchangeable.

Initially, an ASIC contained dedicated logic values and could not be reconfigured [22]. Nowadays, such an abbreviation is typically used in reference to devices developed by a single company and designed as a specific-performance circuit [23].

The first programmable devices were introduced in the 1970s. Such circuits, called as *programmable logic devices (PLDs)*, were built as a fixed array of AND (OR) functions driving a programmable array of OR (AND) functions [22]. Initially, they were used to implement simple combinational logic, however, later registered and tri-state outputs were added.

Programmable logic devices can be generally divided into three main groups: *programmable read-only memory, PROM*, *programmable logic array, PLA* and *programmable array logic, PAL*. The main differences between those devices relay in their structures (fixed/programmable array of AND/OR gates), speed and configurability [22, 35]. A sample structure of a PLD (PAL) is presented in Fig. 9.1 [35].

An extended version of PLDs was presented in early 1980s. A typical *complex PLD (CPLD)* simply consists small PLD blocks (also called *simple programmable logic device, SPLD*) linked to the interconnect arrays. The manufacturing technology of CPLDs depends on the vendor, however, the idea is the same: all the programmable macrocells (and SPLDs) share the common interconnect array.

© Springer International Publishing Switzerland 2017

R. Wiśniewski, *Prototyping of Concurrent Control Systems Implemented in FPGA Devices*, Advances in Industrial Control,
DOI 10.1007/978-3-319-45811-3_9

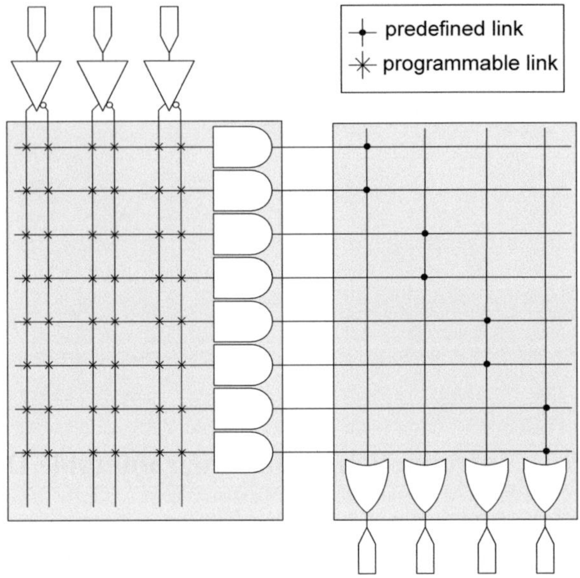

Fig. 9.1 Sample structure of the PLD

The typical macrocell contains AND-OR matrix. Usually, it is a simple PLD, such as PAL or PLA device. Additionally, a macrocell consists of programmable flip-flops, and logic elements (for example multiplexers, XOR gates). Since macrocells surround the interconnect array, they are fully configurable. Unfortunately, it is not always possible to implement large and complex functions, thus they ought to be decomposed [11, 19]. Figure 9.2 shows the typical structure of a CPLD.

Field programmable gate arrays (FPGAs) were introduced in mid-1980s. A typical FPGA consists of a matrix of programmable logic blocks (note, that the name of such a block differs depending on the device vendor). Their structure (logic blocks surrounded by "a sea" of programmable interconnections [22]), performance and very high configurability (even with a possibility of partial reconfiguration) resulted in application in various aspects of human life, such as medicine [28, 31], cryptology [9, 13, 15, 26, 27], aerospace engineering [29], image processing [7, 10, 18], reconfigurable computing [17], measurement [25] and—of course—in concurrent control systems [3, 8, 12, 24, 34].

Next section deals with FPGAs in more details. We shall show the structure of such devices (mainly based on the Xilinx *Virtex-5* family). Moreover, the idea of partial reconfiguration (static and dynamic) of an FPGA will be presented and explained.

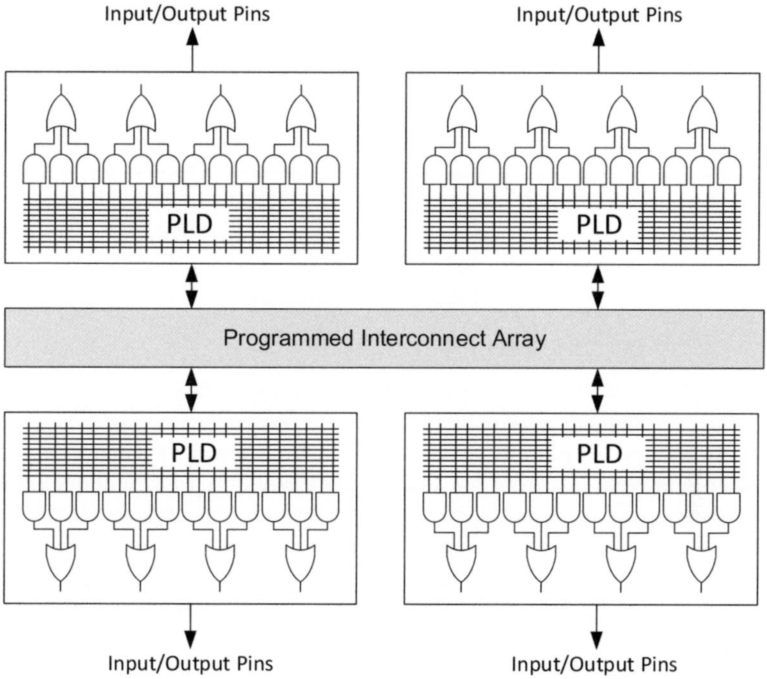

Fig. 9.2 Structure of CPLD

9.2 Field Programmable Gate Arrays

Figure 9.3 illustrates the general structure of the FPGA. Note, that there are different names for internal elements of the FPGA, since each vendor usually uses own notations [1, 2, 20, 22, 35, 39]. Therefore, all the references in this book concern Xilinx FPGAs [39] (more precisely: most descriptions and resource analysis are based on the *Virtex-5* family [44], which is used as a referred device for synthesis, implementation and partial reconfiguration of concurrent control systems).

The main "heart" of an FPGA is the matrix of *configurable logic blocks* (*CLBs*). The CLBs form an array of rows and columns as illustrated in Fig. 9.3. Such elements are connected via *programmable interconnects* [22]. Furthermore, additional storage blocks, called *Block RAMs (BRAMs)*, are located in the device. The communication with the environment is assured by an *Input and Output (I/O) blocks*.

The description of the FPGA is structured as follows: initially, brief overview of BRAMs and I/O blocks is given. Then, we shall show the CLB structure in more details.

Block RAMs (also called *dedicated memory blocks*) are a portion of storage elements in the FPGA. Their main advantage is configurability and size. For example, BRAM in *Virtex-5* FPGA is able to store up to 36 K bits of data. Furthermore, each

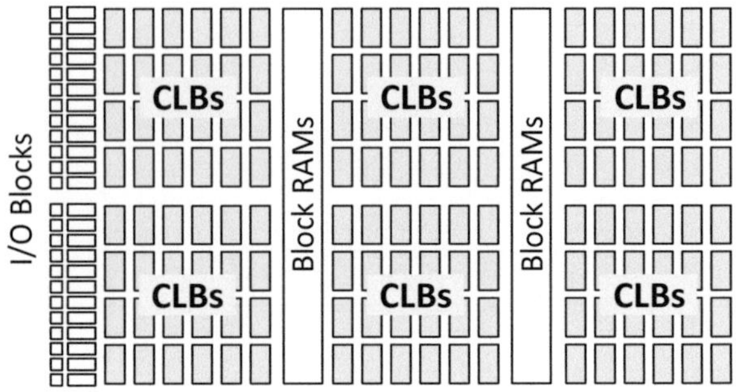

Fig. 9.3 General structure of an FPGA

memory block can be configured independently, composing of various numbers of microoperations and microinstructions, depending on the designer's needs. Typically, BRAMs are located in columns, across the device (cf. Fig. 9.3). Therefore, they can be accessed relatively fast from the combinational logic (CLBs) without additional delays. Note, that Xilinx block RAMs are synchronous, thus clock signal ought to be delivered.

The connectivity between the FPGA and other elements of the prototyped system is assured by an *Input/Output blocks* (also called as an *I/O logic*). An essential role of such blocks is to assure the proper power supply standard [39]. IOBs are organized into banks. Each bank can be configured independently, thus there is a possibility of different I/O standards usage, depending on the standard or voltage needs.

An early CLB contained a simple multiplexer, a flip-flop and a 3-input *look-up table (LUT)*. In a fact, LUT was able to perform any combinational function restricted up to three inputs and one output (more complicated functions ought to be decomposed). Furthermore, each LUT was connected with a storage element (flip-flop) and multiplexer. Therefore, sequential logic could be realized as well.

Later, 4-input LUTs were introduced. These elements provide a base for popular Xilinx families, such as *Spartan-3E* or *Virtex-II Pro* [42, 43]. In opposite to the early versions, the structure of CLB was changed. Each block was divided into four interconnected elements, called *slices*. The slices were additionally grouped in pairs, organized as columns with independent carry chains [42].

Figure 9.4 presents a simplified structure of a slice that contains two 4-input look-up tables. Each LUT is able to perform any logic function up to four input variables. Outputs of both LUTs are connected to the additional carry and arithmetic logic, multiplexers, and two flip-flops. The registers can be programmed as a typical *D-type* flip-flop (with either synchronous or asynchronous set and reset signals) or as a transparent latch [42]. The arithmetic logic contains various logic gates dedicated for arithmetic operations, while carry chain provides signals from the previous and to the next slice in the column.

Another modifications to the CLBs were made starting from the *5-series* FPGAs. Each CLB was divided into two slices. The first one is located in the bottom while the remaining one resides in the top of the block. Both slices are linked to the common

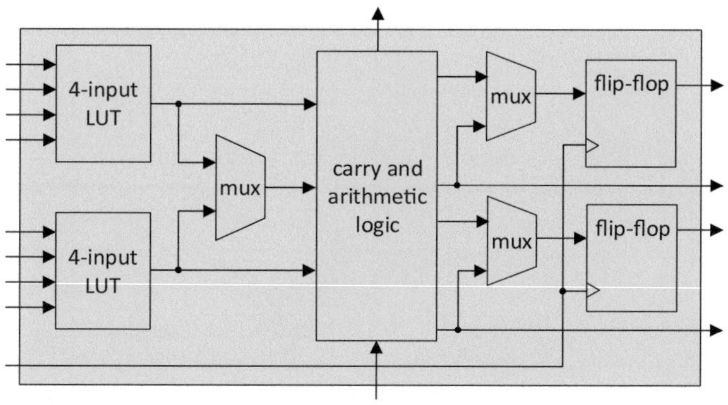

Fig. 9.4 Simplified structure of a slice (4-input LUTs)

programmable interconnections matrix. Additionally, slices are organized in rows and columns. Let us describe such a localization in more details, since it is essential for the understanding of partial reconfiguration of the device.

Figure 9.5 illustrates the organization of slices and CLBs in the *Virtex-5* FPGA. Note, that position of each slice in the device is denoted by coordinates "X" and "Y". Starting from the left-down piece of the FPGA, the first slice (in the bottom of a CLB) is denoted by "X0Y0". The upper slice in this block has the same "Y" coordinate, while "X" is increased by one. The coordinates of the remaining slices are calculated in the similar way [44]. For example, slices located in the block on the right from the one already described are denoted by "X2Y0" (the bottom slice) and "X3Y0" (the upper slice). Slices of the upper CLB have the coordinates "X0Y1" (bottom one), "X1Y1" (upper), and so on.

Note, that there are two carry chains that intersect each CLB. Such chains connect all the slices located in the same column, that is, slices sharing common "X" coordinate.

Figure 9.6 shows a simplified structure of a slice of the *Virtex-5* FPGA. The typical slice consists of

- four 6-input look-up tables,
- four storage elements (configured as a *D-type* flip-flops or latches),
- multiplexers,
- carry and arithmetic logic.

Note that there are two types of slices in the *Virtex-5* FPGA. The typical one is called *SliceL*. However, some slices support additional functions, such as storing data (as a distributed *random-access memory, RAM*) or as a shift registers. Those slices are denoted as *SliceM*.

There are six independent inputs and two independent outputs for each LUT. The performance of both outputs depends on the type of realized logic functions. If the look-up table performs a combinational logic of all six inputs, only one output

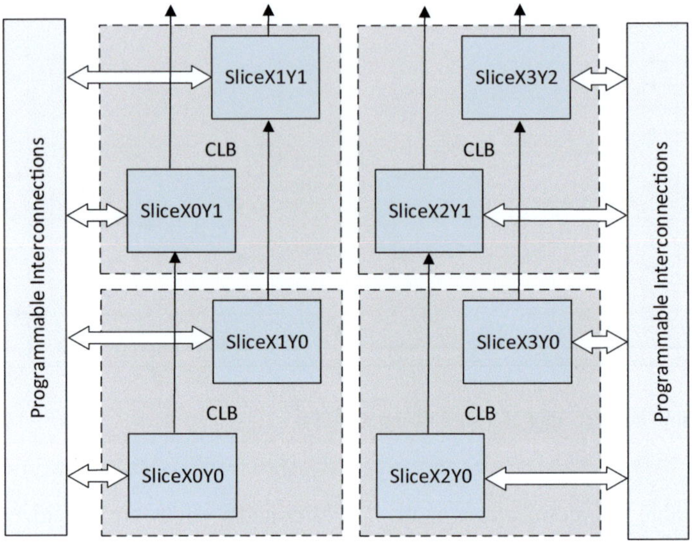

Fig. 9.5 Organization of slices and CLBs in the *Virtex-5* FPGA

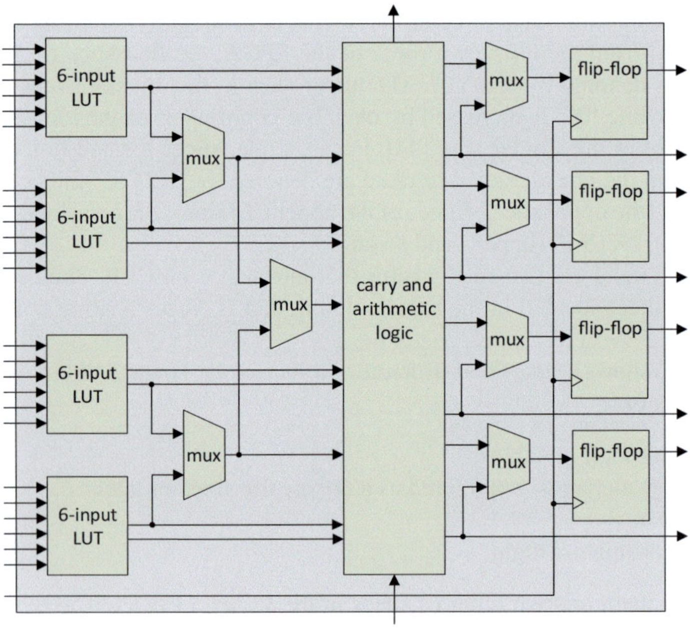

Fig. 9.6 Simplified structure of a slice in the *Virtex-5* FPGA (6-input LUTs)

is used. However, there is a possibility to implement two arbitrary five-input logic functions sharing common variables in the single LUT. In such a case, both outputs are used [44].

9.3 Implementation of Concurrent Controllers in FPGA

This section deals with the implementation of the concurrent control systems in an FPGA device. It is assumed that Reader is familiar with the prototyping flow of concurrent controllers (cf. Chap. 7) and FPGA preliminaries (introduced in the previous section).

Note, that meaning of the terms such as *logic synthesis*, *logic implementation* may be slightly different, depending on the vendor of the destination FPGA and the applied software. In our considerations all the descriptions regard Xilinx FPGAs (*Virtex-5 family*) and *Xilinx ISE Design Suite* (version 14.7). Please also note, that prototyping flows for other vendors (such as Altera or Atmel) or with the use of different software (like *Vivado Design Suite*) may be slightly different, however the implementation idea of concurrent control systems (as well as the concept of partial reconfiguration of such controllers) is exactly the same.

Figure 9.7 shows the prototyping flow of the concurrent control system (left) with enumerated stages of the implementation in an FPGA (right). Based on the description of the system in hardware languages (cf. Chap. 8), the design is logically synthesized and implemented. Then, the programming data (called *bit-stream*) are generated and the FPGA can be physically configured. Let us briefly present each of the implementation stages.

Logic synthesis of the prototyped concurrent control system is a process that converts the controller into the specific *net-list* files. In other words: synthesis transforms the design into a gate-level representation [33].

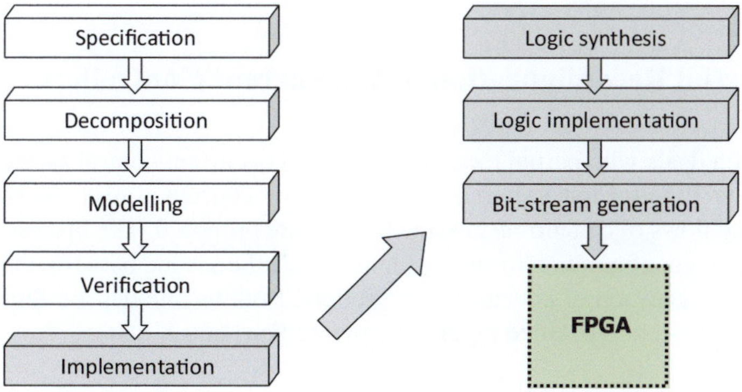

Fig. 9.7 The prototyping and implementation flows of the concurrent control system

The resulting net-lists may be written in various formats. In our considerations we shall use an *native generic circuit, NGC* format. Moreover, particular modules of the prototyped system can be synthesized separately, producing independent NGC files. We shall use this property during the partial reconfiguration of the concurrent control systems (cf. Sect. 9.4).

The synthesis process permits estimation of the required hardware resources that are needed in order to implement the system (or particular module). Such values are simply expressed in utilized *slice LUTs* and slice *flip-flops* of the destination FPGA.

Logic implementation consists of three main substages, called *translate*, *map* and *place and route*. The first one merges all the net-lists obtained during the synthesis process and produces the logical description of the system in a form of *primitives*, that is, hardware elements of the destination device. At this step, the *constraint data* ought to be provided. Such data contain information about the placement of particular input/output signals, definition of I/O standards and voltage, timing of clock signal(s), and so on [40]. Typically, constraints are specified in a *user constraint file, UCF*. The second substage of the logic implementation distributes the prototyped system over the target FPGA. Finally, the placement and routing of the design is performed. At this step, the timing constraints are verified, in order to satisfy the required user expectations [40].

Bit-stream generation simply produces the portion of data, that is used for configuration of the destination device. In the traditional prototyping step (without partial reconfiguration) such a file contains the full description of the FPGA. Note, that the size of the typical bit-stream file exceeds one million bits. For example, the FPGA *XC5VLX50* (*Virtex-5 family*, a part of *ML501 Evaluation and Development Platform*) requires 1 569 676 bits to configure the device. Such a portion of data ought to be sent to the FPGA each time the new version of the concurrent control system is prepared by designers. However, there is a technique that allows reducing the size of the configuration file. Furthermore, there is even a possibility to exchange a part of the device, while the rest of the FPGA remains untouched. The next sections proposes *partial reconfiguration* of the concurrent control systems implemented in the FPGA.

9.4 Partial Reconfiguration of Concurrent Controllers

This section deals with partial reconfiguration of concurrent control systems implemented in an FPGA. First, a short overview of such a concept is given. Next, two new prototyping flows of concurrent control systems are proposed. The first one includes *static* partial reconfiguration of such systems, while the second bases on the *dynamic* partial reconfiguration of concurrent controllers. Both techniques are illustrated by examples of systems specified by the interpreted Petri nets.

9.4.1 The Idea of Partial Reconfiguration of an FPGA

Partial reconfiguration of field programmable gate arrays is relatively new concept. However, most of popular FPGAs that are currently available on the marked support partial reconfiguration (cf. [4, 41]). In opposite to the traditional designing flow, the whole version of the prototyped system is usually implemented only once. Further changes of the FPGA restricts to the particular area of the device, previously selected by the designer.

Let us illustrate the concept of the partial reconfiguration. Figure 9.8 shows a piece of the FPGA device. Assume, that a part of the implemented system (denoted by *blocks to be reconfigured*) ought to be replaced by a newer version. Partial reconfiguration permits exchanging only this area of the device, while the rest of the implemented design (marked as *other blocks used by the prototyped system*) remains untouched.

Looking from the perspective of the design functionality, partial reconfiguration can be split into two main parts [35]:

- *Static partial reconfiguration*—the FPGA is stopped during the reconfiguration process. While the partial portion of data is sent into the device, the rest of the FPGA is in the shutdown mode (not active). The device is brought up after the partial configuration is completed.
- *Dynamic partial reconfiguration*—the FPGA is active (not stopped) during the reconfiguration process. While the part of the device is being reconfigured, the rest of the FPGA is still running. In other words, dynamic partial reconfiguration permits configuration of the FPGA without stopping of the device.

Next sections present the prototyping flows for both: static and dynamic partial reconfigurations of concurrent control systems implemented in the FPGA device. The proposed methods are illustrated by examples of real-life controllers.

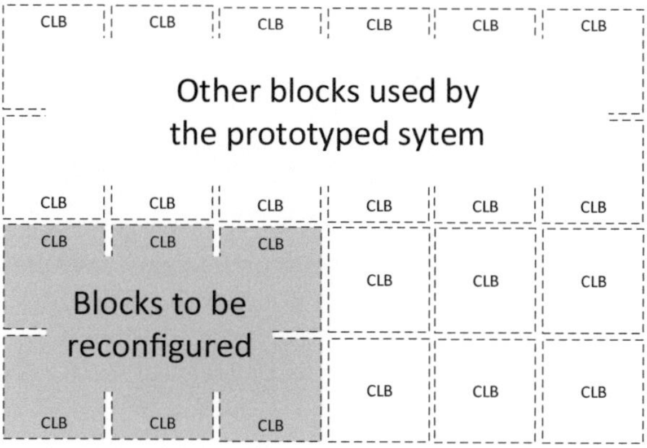

Fig. 9.8 The idea of the partial reconfiguration of the FPGA

Note, that proposed ideas rely on the specification of the system as an interpreted Petri net and the further decomposition of the controller into separate modules. Other reconfiguration techniques of concurrent controllers can be found in [6, 8, 14, 16, 21, 30, 36, 38].

9.4.2 Partial Reconfiguration of Concurrent Systems

Figure 9.9 shows the idea of the proposed partial reconfiguration of a concurrent control system. It is assumed, that there are at least two versions of the prototyped system. The first one (main version) is implemented in the FPGA only once. Other versions (called *contexts*) are used in order to exchange the part of the device.

The presented technique strongly relies on the decomposition of the controller. Indeed, *contexts* are just different versions of the decomposed module that is to be replaced. We shall denote such a module as a *reconfigurable module*. In the other words, *reconfigurable module* is a module of the prototyped system, that contains at least two (or more) contexts.

Figure 9.10 shows a sample concurrent control system, decomposed into four modules. Assume, that *Module C* is a reconfigurable module and there are three contexts. The first context is implemented during the initial configuration of the FPGA. After that, the device is partially reconfigured with the use of contexts of *Module C*. Note, that it is possible to return back to the initial version of the system, because the first context of the module is also available for partial reconfiguration.

The part of the system that remains unchanged shall be called *core*. Simply, the core refers to all the modules that are not partially reconfigured. In case of the system presented in Fig. 9.10, modules *A*, *B* and *D* form a core, since they are not reconfigurable modules.

Note, that the number of inputs and outputs of the reconfigurable module in all its contexts should be constant. Thus, all the input and output signals that are utilized by any version of this particular module ought to be specified in all the contexts. We shall explain such a situation later, by an example.

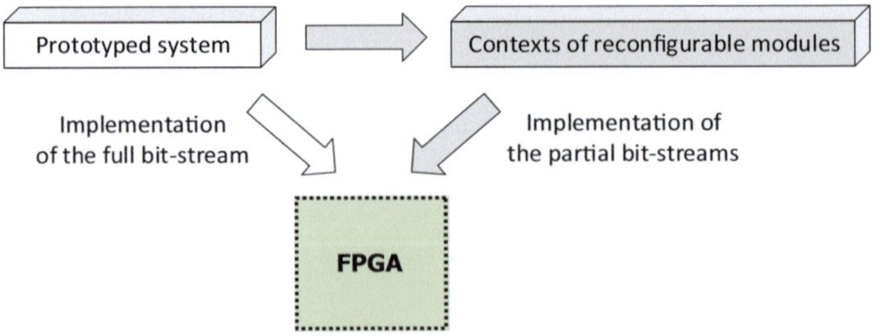

Fig. 9.9 The idea of the partial reconfiguration of the concurrent control system

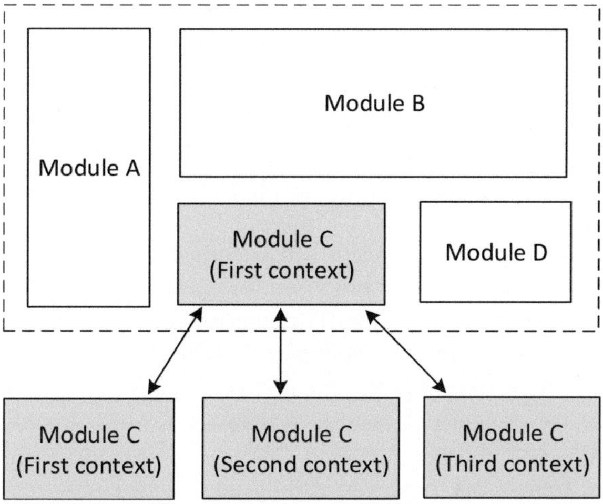

Fig. 9.10 Sample controller with reconfigurable *Module C* and three contexts

Depending on the type of the partial reconfiguration, the design flow is slightly different. Let us analyze both, static and dynamic partial reconfiguration prototyping flows of concurrent control systems in more details.

9.4.3 Static Partial Reconfiguration

The proposed prototyping flow of concurrent control systems for further static partial reconfiguration in the FPGA includes the following steps:

1. Specification of the first version of the concurrent control system (concurrent controller) by an interpreted Petri net.
2. Decomposition and synchronization of the concurrent controller.
3. Modelling of the first version of the concurrent controller.
4. Verification of the first version of the concurrent controller.
5. Preparing of the additional contexts for each of reconfigurable modules. For each new context:
   (a) Specification of the new context of the reconfigurable module.
   (b) Modelling of the new context of the reconfigurable module.
   (c) Verification of the concurrent controller with the new context.
   (d) Validation of the previous contexts.
6. Implementation of the concurrent controller with additional contexts as a system for further partial reconfiguration:
   (a) Logical synthesis of the *core* of the concurrent controller.
   (b) Logical synthesis of each context of each reconfigurable module.

(c) Assignment of the area of the FPGA for reconfigurable modules.
(d) Logical implementation of the *core* together with all contexts.
(e) Generation of bit-streams for each versions of the controller.
(f) Physical implementation of the FPGA by the *core* and by the first context (contexts) of reconfigurable modules.

7. Partial reconfiguration of the FPGA with the use of selected contexts.

The initial four steps are executed exactly in the same manner as in case of the prototyping flow proposed for integrated concurrent control systems in Sect. 7.2. Clearly, the first version of the controller includes the core of the system. Furthermore, it consists of the first context of modules, that are to be partially reconfigured.

The fifth step of the above flow is an essence of the static partial reconfiguration of the concurrent control systems. For each reconfigurable module, new context (or contexts) are prepared. There are no restrictions regarding number of places or transitions, however, it is important to keep proper synchronization with other modules. A new context may be easily specified by a Moore automaton and described in HDLs, according to the guidelines shown in Chap. 8. Next, the context is verified together with the remaining logic (*core*) of the prototyped system (new context simply replaces the previous version of modified module). This procedure is repeated for each reconfigurable module and for each new context. Note, that adding a new version of the reconfigurable module requires validation of its previous contexts. For example, additional input/output signals have to be updated in all the previous versions of the reconfigurable module.

Once the contexts for reconfigurable modules are ready, the whole system ought to be prepared for further partial reconfiguration. Initially, the *core* of the system is synthesized. Such a process requires to specify the reconfigurable modules as *Black Boxes* [41], that is, empty modules with input and output ports declaration. The logic of reconfigurable modules are synthesized separately. Therefore, for each context of each module an independent *net-list* file in an *NGC* format is achieved (cf. Sect. 9.3).

The logical implementation of the concurrent control system for further partial reconfiguration mainly relies on the assignment of the FPGA area to be used by reconfigurable modules. The selection is performed for each reconfigurable module separately. Figure 9.11 shows a sample assignment of the logic elements to the reconfigurable module *module_C*. In this particular example, three pairs of *SliceL* and *SliceM* blocks are utilized. Note that assignment of the area for reconfigurable modules directly influences the size of the partial bit-stream file.

Further logical implementation and generation of the bit-streams are fully automated. As a result, the set of configurable files are produced. Indeed, for each context of each module two bit-streams are generated. The first one contains information about the whole system (full bit-stream), while the second one is used for further partial reconfiguration (reduced bit-stream). It means, that any version of the concurrent control system can be initially implemented in the FPGA.

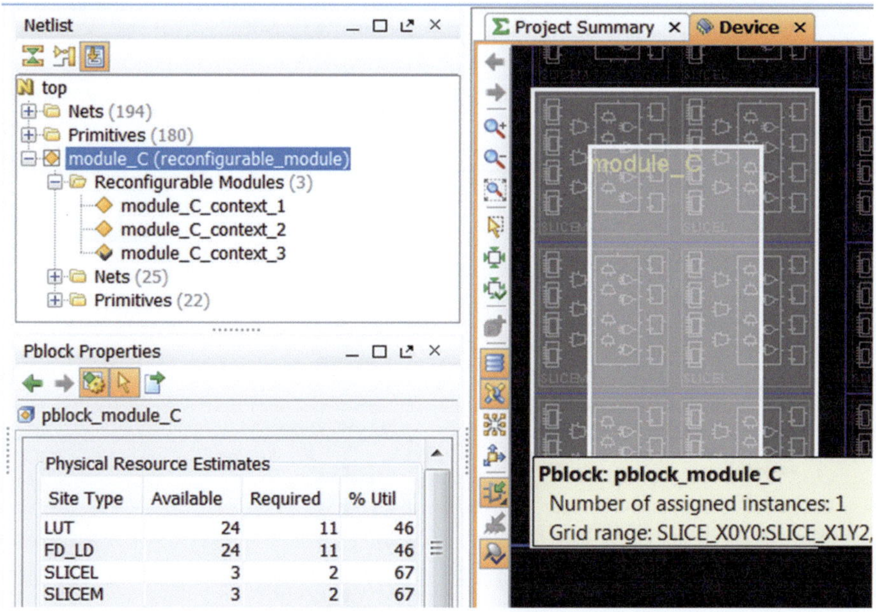

Fig. 9.11 Assignment of the FPGA area for reconfigurable module

Partial reconfiguration of the already implemented system can be done at any time. However, it is recommended to reset the concurrent controller to avoid malfunctions [41]. Since other contexts of reconfigurable modules contain various internal states, combinational and sequential logic, it may lead to unexpected behavior of the system. Therefore, such modules should be reset to the initial state.

Let us now show the static partial reconfiguration of concurrent control systems by examples. Recall the milling machine presented in Fig. 2.3 and specified by the interpreted Petri net PN_2 (Fig. 2.4). Decomposition and synchronization of the system lead to four state machine components shown in Fig. 9.12.

The third component $S_3=\{NOP_2, p_7, p_8, p_9, p_{10}, p_{11}, p_{12}, p_{13}\}$ is responsible for cutting the proper shape from the wooden plank. At the initial specification, the square is cutout by moving the drill in a given sequence. Figure 9.13 introduces a new version for the third module. The presented context is responsible for cutting out the “U-shape” from the plank. After immersion into the wood (y_4), the drill moves to the right (y_5), until reaching the position indicated by sensor x_{15}. Then, the drill goes down, to the right (y_5), to the top (y_8) and once more to the right. The remaining positions are signalized by sensors x_{17}, x_{16} and x_4, respectively. The subsequent drill movement is exactly the same as in the first context. It goes down (y_6), to the left (y_7) and to the top (y_8), according to sensors x_5, x_6, and x_7. Finally, the drill moves up (y_9) to its initial position.

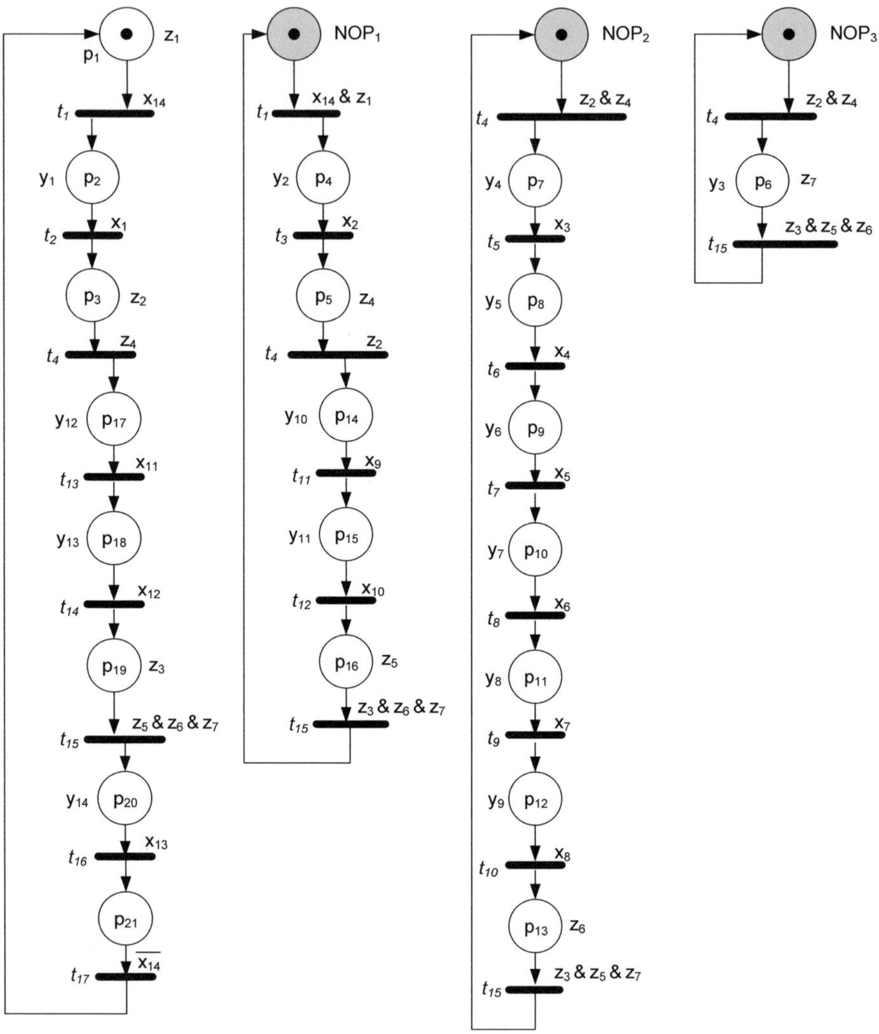

Fig. 9.12 Decomposed and synchronized net PN_2

The second context of the third component contains twelve places and twelve transitions. Note, that shared transitions t_4 and t_{15} remain unchanged. Moreover, synchronization signal z_6 is now generated by place p_{32}.

The movement of the drill is coordinated by three sensors x_{15}, x_{16} and x_{17} that were not used in the previous version. Therefore, those inputs ought to be added to the specification of the first context.

The logic synthesis and implementation of the described concurrent control system result in four configuration files:

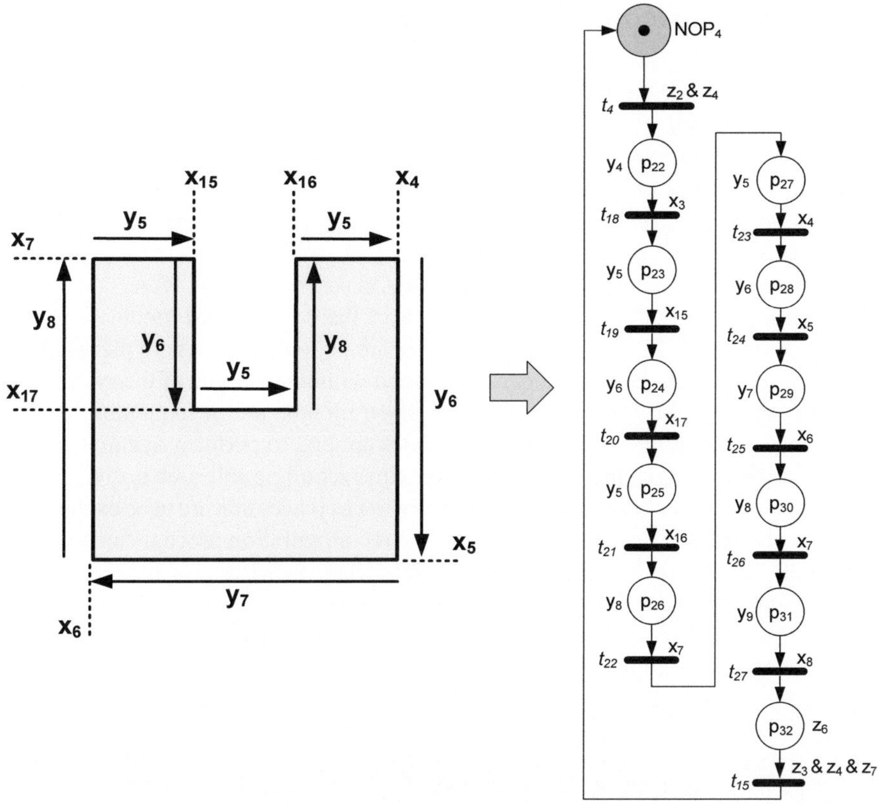

Fig. 9.13 Second context of the reconfigurable module of the milling machine

- Full bit stream for the first context.
- Full bit-stream for the second context.
- Partial bit-stream for the first context.
- Partial bit-stream for the second context.

Finally, the system is physically implemented in an FPGA with the use of one of the full bit-streams. Assume, that first context is downloaded to the device. The milling machine cutouts the squares from a given wooden plank. Further reconfiguration with the partial bit-stream for the second context changes the functionality of the controller. Now, the system cutouts the "U-shape" from the wood. Such a process is performed until the FPGA is partially reconfigured again, this time with the first context. Of course, it is possible to add some other contexts. For example, the milling machine can be used for cutting of the "L-shape" [37] or any other required shapes.

9.4.4 Dynamic Partial Reconfiguration

The idea of dynamic partial reconfiguration is different from the static one. The system is still running, while the part of the design is exchanged. Therefore, there are additional conditions and restrictions that ought to be fulfilled. Let us analyze such requirements in more details.

Looking from the functionality of the prototyped system, the main problem that ought to be solved is the selection of a reconfigurable area. Of course, it is possible to reconfigure whole modules, however it does not make much sense. Note, that in case of a controller specified by a Petri net, places of the decomposed module intersect all the markings (cf. Chap. 5). Therefore, there are no markings, where places of the particular SMC can be dynamically reconfigured without stopping of the system. On the other hand, all the places of the decomposed module are in sequential relation (cf. Chap. 5). This property can be successfully applied to perform dynamic partial reconfiguration. Simply, the module that contains reconfigurable area is divided into two parts: *dynamic* and *static*. The first one consists of places that are to be exchanged, while the second part remains unchanged. The reconfiguration process can be safely performed, because both parts are sequential. Thus, the first part is reconfigured while the controller executes actions in the *static* part.

Due to the requirements of the dynamic partial reconfiguration, both parts must be described as additional modules. Moreover, additional synchronization signals ought to be supplied to the system. Note, that it is assumed that the part of only one of decomposed modules is reconfigured dynamically.

The proposed prototyping flow of concurrent control systems for further dynamic partial reconfiguration in the FPGA includes the following steps:

1. Specification of the first version of the concurrent control system (concurrent controller) by an interpreted Petri net.
2. Decomposition and synchronization of the concurrent controller.
3. Preparing of the reconfigurable area in one of the decomposed modules:
    (a) Selection of the reconfigurable area.
    (b) Supplementation of the system by reconfiguration signals:
        - *Reconfiguration request* (denoted by *Rec* or *rec*),
        - *Reconfiguration allowed* (denoted by *Ral* or *ral*).
    (c) Splitting of the module into *static* and *dynamic* submodules (parts) according to the performed selection.
    (d) Configuration of the *Reset* signal in the *dynamic* submodule.
4. Modelling of the first version of the concurrent controller.
5. Verification of the first version of the concurrent controller.
6. Preparing of the additional contexts for the *dynamic* submodule:
    (a) Specification of the new context of the *dynamic* submodule.
    (b) Modelling of the new context of the *dynamic* submodule.

   (c) Verification of the concurrent controller with the new context.
   (d) Validation of the previous contexts.

7. Implementation of the concurrent controller with additional contexts as a system for further partial reconfiguration:

   (a) Logical synthesis of the *core* of the concurrent controller.
   (b) Logical synthesis of each context of each *dynamic* submodule.
   (c) Assignment of the area of the FPGA for *dynamic* submodule.
   (d) Logical implementation of the *core* together with all contexts.
   (e) Generation of bit-streams for each versions of the controller.
   (f) Physically implementation of the FPGA by the *core* and by the first context (contexts) of *dynamic* submodule.

8. Partial reconfiguration of the FPGA with the use of selected contexts.

Let us describe the above prototyping flow in more details. Additionally, we shall illustrate it by an example. The first two steps are executed in exactly the same manner as in the case of the traditional prototyping methodology. The concurrent controller is specified by an interpreted Petri net which is decomposed with the application of one of the methods proposed in Chap. 6.

Figure 9.14 (left) shows an exemplary concurrent control system specified by an interpreted Petri net PN_8. The net contains six places and four transitions. There are six outputs $Y=\{y_1, \ldots, y_6\}$, each of them executed in a particular place. The decomposition (and further synchronization) of PN_8 results in two SMCs, as it is presented in Fig. 9.14 (right). The first module contains places $S_1=\{p_1, p_3, p_4, p_5\}$, while the second consists of $S_2 = \{p_2, NOP_1, p_6\}$.

Third step is the most important stage of the whole designing flow since it strictly influences the further dynamic reconfiguration of the concurrent system. The reconfigurable area (also denoted a as *dynamic part*) should be chosen very carefully with paying special attention to the other (concurrent) modules to the one that contains places to be dynamically exchanged. It is strongly recommended to select the *dynamic* part between the transitions that are shared by other decomposed modules. Clearly, all the remaining places of the selected module form the *static* part (do not confuse with the *static reconfiguration* described in the previous section). Note, that the analysis of the concurrency and sequentiality properties (shown in Chap. 5) may help greatly in the selection of the *dynamic* part.

In the presented example, we shall dynamically reconfigure a fragment of the first module (S_1). Let us form the *dynamic* part from places p_4 and p_5. Thus, the *static* part consists of places p_1 and p_3. Note, that input and output transitions of the *dynamic* part are shared by both SMCs.

Once the reconfiguration area of the concurrent controller is selected, the additional reconfiguration signals ought to be supplemented. The first one, *reconfiguration request* signal (denoted by *Rec* or *rec*) should be delivered to all of the decomposed SMCs. It assures proper functionality of the controller during the dynamic reconfiguration process. Loosely speaking, *reconfiguration request* prevents the system from entering the places that are concurrent to the reconfigurable area. In the

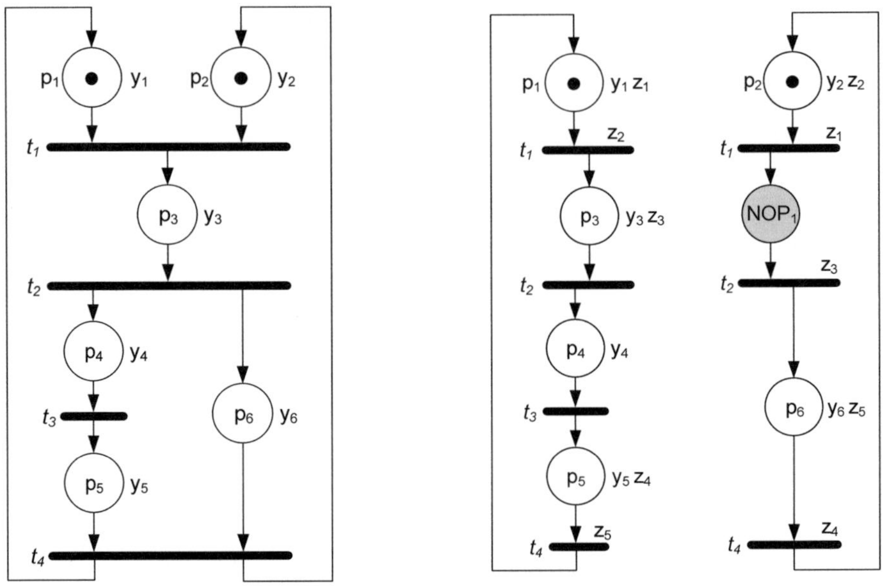

Fig. 9.14 Interpreted Petri net PN_8 (*left*), decomposed and synchronized (*right*)

presented example, *rec* is added to both modules as an additional condition for transition t_2 (Fig. 9.15).

The dynamic reconfiguration can be safely performed as soon as the *reconfiguration allowed* signal is active. Such a signal is set in all places that belong to the *static* part.

The general idea of splitting the module into *static* and *dynamic* parts is shown in Fig. 9.16. Note, that *Reset* signal is additionally gated by both reconfiguration signals (*Rec* and *Ral*) in order to satisfy the *Xilinx* requirements (*Reset* signal should be applied to the *dynamic* part of the system once the reconfiguration finishes [41]).

Splitting of the module into *static* and *dynamic* submodules are relatively easy. The process is very similar to the state partitions of an FSM shown in [5]. Both submodules (*static* and *dynamic*) are supplemented by *nonoperational places*. Simply, the first one (NOP_S) replaces the dynamic part in the *static* submodule, while the second one (NOP_S) exchanges the static part in the *dynamic* submodule. Note, that *static* submodule is a part of *core*, since new contexts are prepared only for *dynamic* submodule.

Figure 9.17 shows the first module (S_1) of decomposed PN_8 split into *static* (upper) and *reconfigurable* submodules (lower). The *static* submodule contains places S_S=$\{p_1, p_3, NOP_R\}$. The additional place NOP_R in this submodule replaces the reconfigurable part (places p_4 and p_5) that are located in the second submodule. The *dynamic* submodule consists of places S_R=$\{NOP_S, p_4, p_5\}$. Place NOP_S replaces the static part of S_1, that is, places p_1 and p_3. The synchronization between both submodules is assured by signals achieved during decomposition of the system. Thus,

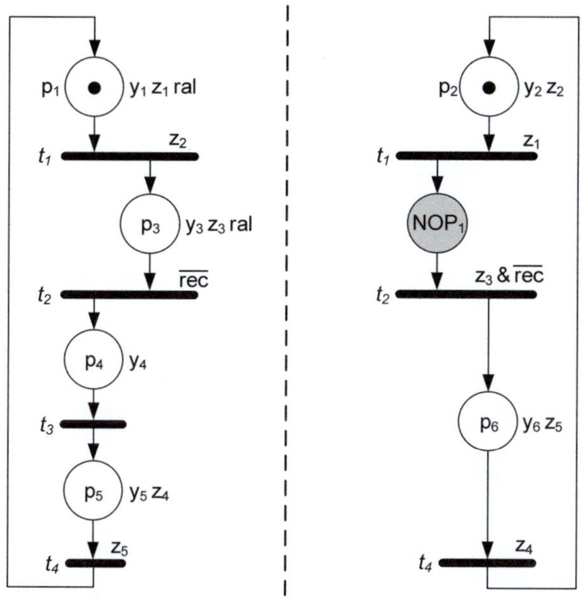

Fig. 9.15 The decomposed net PN_8 supplemented by reconfiguration signals

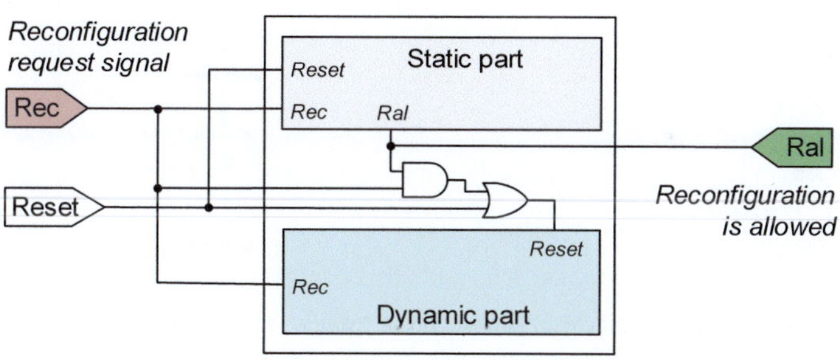

Fig. 9.16 The idea of splitting the module into *static* and *dynamic* submodules

z_3 is added as a condition to synchronize t_2 in the *dynamic* submodule, while z_4 (together with z_5) synchronizes t_4 in the *static* submodule.

According to the *Xilinx* outlines [32, 41], the reconfigurable part of the system ought to be reset after the partial programming. Therefore, *Reset* in the *dynamic* submodule is additionally gated by the reconfiguration signals, as it is shown in Fig. 9.16.

At the next stage of the proposed prototyping flow, the whole system is modelled in the hardware description languages. Note, module S_1 is split into two submodules: *static* and *dynamic*. Both of them are described according to the guidelines shown in Chap. 8, as it is shown in Listing 9.1.

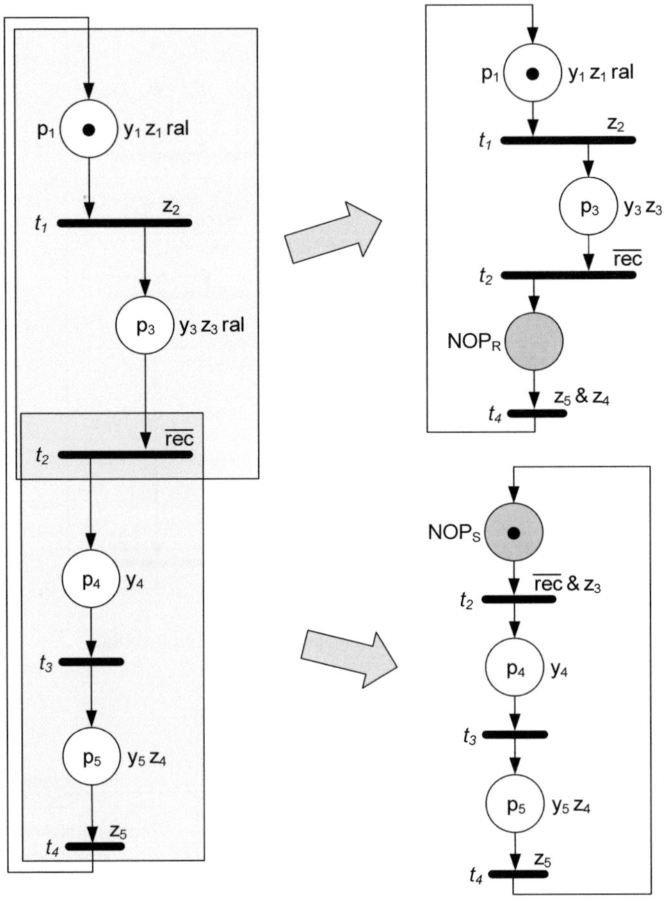

Fig. 9.17 Splitting of the module S_1 into *static* and *dynamic* submodules

Listing 9.1 Description of the module S_1 (PN_8) with *static* and *dynamic* submodules

```
module S1(y1,y3,y4,y5,z1,z3,z4,ral,clk,reset,z2,z5,rec);
    output y1,y3,y4,y5,z1,z3,z4,ral;
    input clk,reset,z2,z5,rec;

    S1_static S1_s (y1,y3,ral,z1,z3,clk,reset,z2,z4,z5,rec);
    S1_dynamic S1_d (y4,y5,z4,clk,reset|(ral&rec),z3,z5,rec);
endmodule

module S1_static(y1,y3,ral,z1,z3,clk,reset,z2,z4,z5,rec);
    output y1,y3,ral,z1,z3;
    input clk,reset,z2,z4,z5,rec;

    reg [1:0] state=0,next=0;
    parameter p1=2'b00, p3=2'b01, nop_r=2'b11;
```

```
    always@(posedge clk or posedge reset)
    if (reset) state<=p1; else state<=next;

    always@(state or z2 or z4 or z5 or rec)
    case (state)
        p1:next<=(z2)?p3:p1;
        p3:next<=(!rec)?nop_r:p3;
        nop_r:next<=(z4 && z5)?p1:nop_r;
        default:next<=p1;
    endcase

    assign {y1,z1}=(state==p1)?2'b11:2'b00;
    assign {y3,z3}=(state==p3)?2'b11:2'b00;
    assign ral=(state==p1 || state==p3)?1'b1:1'b0;
endmodule

module S1_dynamic(y4,y5,z4,clk,reset,z3,z5,rec);
    output y4,y5,z4;
    input clk,reset,z3,z5,rec;

    reg [1:0] state=0,next=0;
    parameter nop_s=2'b00, p4=2'b01, p5=2'b11;

    always@(posedge clk or posedge reset)
    if (reset) state<=nop_s; else state<=next;

    always@(state or z3 or z5 or rec)
    case (state)
        nop_s:next<=(z3 && !rec)?p4:nop_s;
        p4:next<=p5;
        p5:next<=(z5)?nop_s:p5;
        default:next<=nop_s;
    endcase

    assign y4=(state==p4)?1'b1:1'b0;
    assign {y5,z4}=(state==p5)?2'b11:2'b00;
endmodule
```

Once the concurrent control system is verified, new contexts of the *dynamic* submodule is prepared. In our example we shall exchange places p_4 and p_5 by three other places: p_7, p_8, and p_9, as it is presented in Fig. 9.18. Moreover, three new output signals are added to the system: y_7, y_8,s and y_9. The new context is verified together with the *core*. Next, validation of the previous contexts should be performed. Note, that outputs y_7, y_8, y_9 ought to be added to the previous context of the *dynamic* submodule.

Finally, the system is implemented, according to the guidelines shown in the previous section. Partial reconfiguration of the *dynamic* submodule is performed dynamically. It means, that *core* is still working, executing other operations of the concurrent control system.

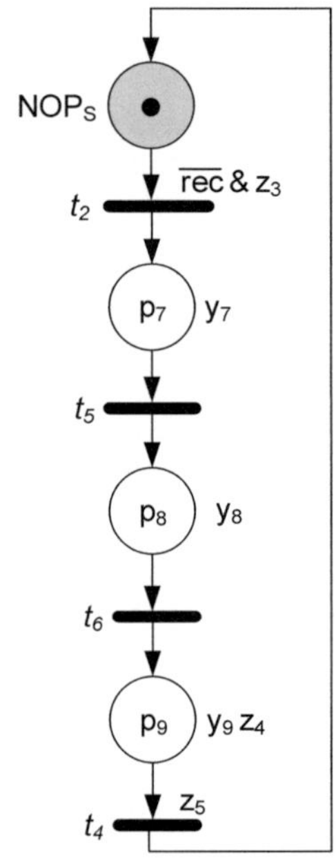

Fig. 9.18 Second context of the *dynamic* submodule

From the technical point of view, the reconfiguration process is handled by two additional signals: *Reconfiguration request (Req)* and *Reconfiguration allowed (Ral)*. The request for the dynamic reconfiguration is provided by the input *Rec* signal, which should be active during the whole configuration process. Active *Ral* signal notifies that the system is ready for downloading of the partial bit-stream.

Let us now emphasize the benefits of dynamic reconfiguration of concurrent control systems by a real-life example. Recall the milling machine presented in the previous section. Now, we shall try to dynamically reconfigure the part of the system that is responsible for cutting shape from the wooden plank (module S_3), as it is denoted in Fig. 9.19. Note, that placing and removing the wooden plank (y_1 and y_{14}, respectively) are performed sequentially to the selected area. We shall take an advantage of this dependency and functionality of the controller will be dynamically reconfigured during the preparation of the wooden plank, which takes a while. Note, that additional reconfiguration signal *rec* is added to the transition t_4 in all four components.

The selected area for further partial reconfiguration includes seven places $\{p_7, \ldots, p_{13}\}$, that together with NOP_S form the *dynamic* submodule of S_3. The *static* submodule consists of only two places: NOP_2 and NOP_R. Splitting of the component S_3 into submodules is illustrated in Fig. 9.20. Note, that although NOP_2 is

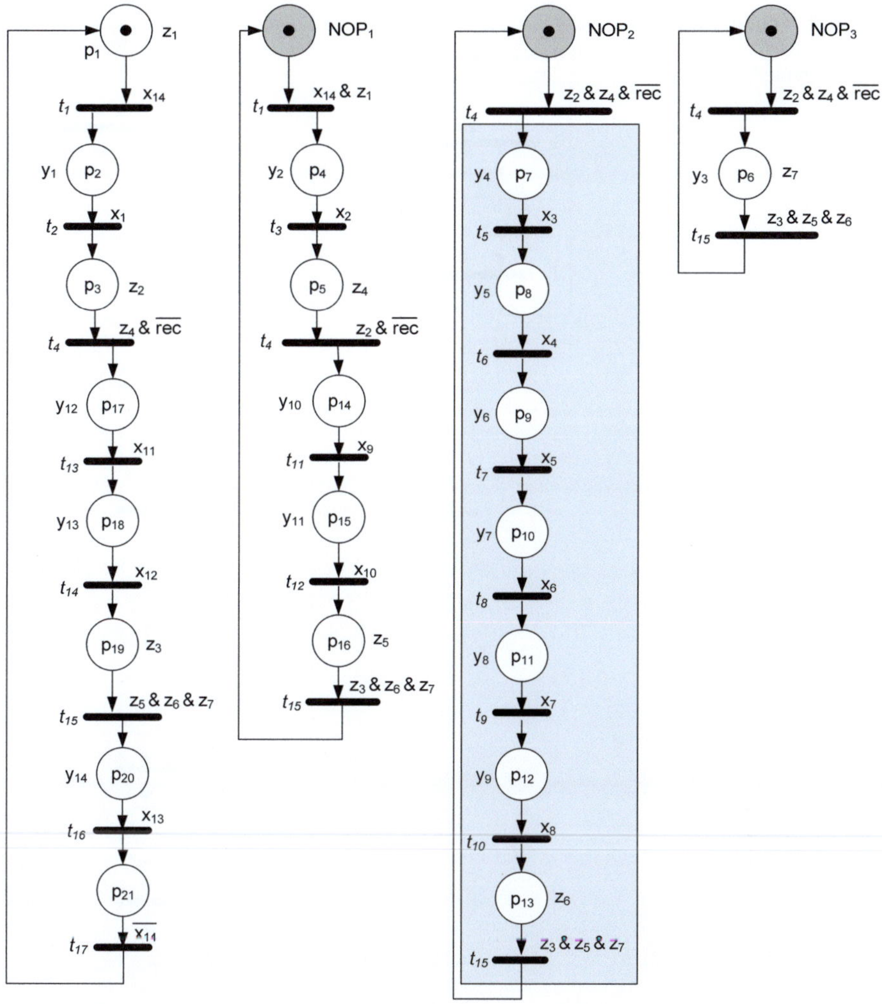

Fig. 9.19 The selected area for partial reconfiguration of the milling machine

a nonoperational place from the point of view of the functionality of the milling machine, now it generates the *Ral* signal in order to coordinate the reconfiguration process. Moreover, an additional synchronization signal s_6 is added to the logical conjunction of signals assigned to the transition t_4 in the *static* submodule.

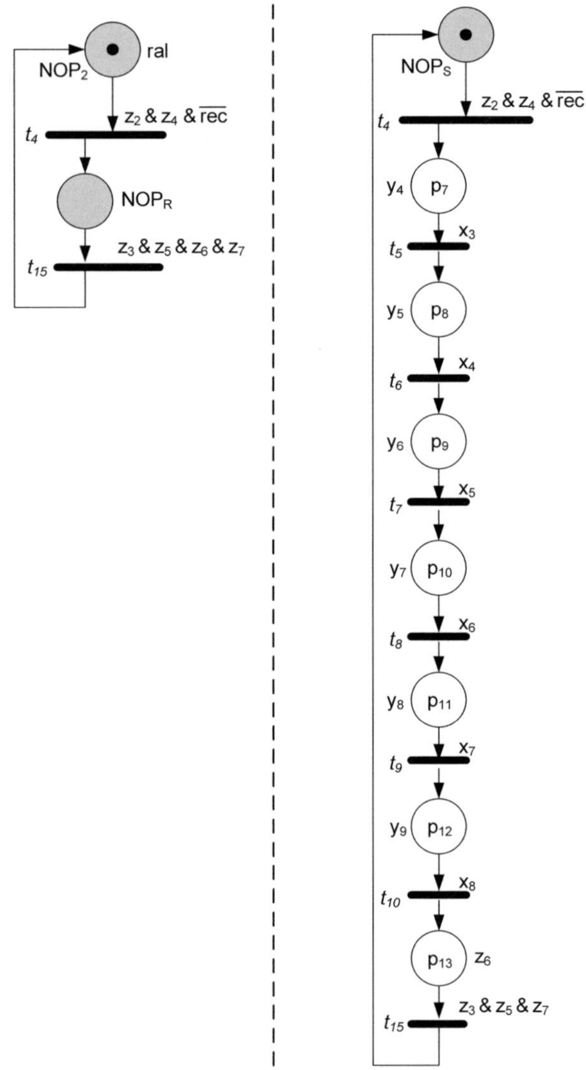

Fig. 9.20 Splitting of S_3 into *static* (*left*) and *dynamic* (*right*) submodules

Once the modelling and verification of the first version of the system is finished, a new context is added. Figure 9.21 presents an alternative version of the *dynamic* submodule. Similarly to the idea shown in the previous section, the new context changes the form that is cutout from the wooden plank into the “U”-shape. Since

Fig. 9.21 New contexts of the *dynamic* submodule (cutting "U" shape)

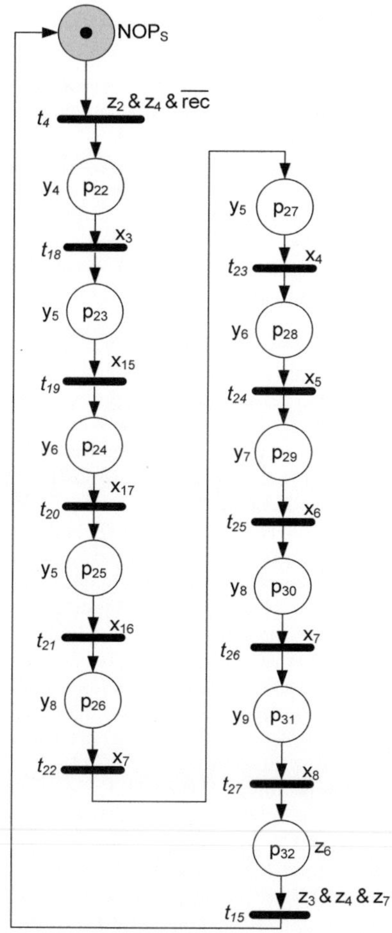

new input and output signals are added to the system, the previous contexts ought to be updated. Further synthesis and logic implementation of the controller results in four bit-streams. Two of them contain full data, while the remaining two—partial data. The partial reconfiguration is performed dynamically, during replacement of the wooden plank. Additional signal *rec* prevents from unexpected behavior of the system if the reconfiguration process takes more time.

References

1. Altera homepage: http://www.altera.com. Accessed 2016-03-04
2. Atmel Programmable Logic homepage: http://www.atmel.com/products/programmable-logic. Accessed 2016-03-04
3. Adamski M, Karatkevich A, Wegrzyn M (eds) (2005) Design of embedded control systems. Springer, New York. ISBN:0-387-23630-9
4. Altera Stratix V FPGAs: Ultimate flexibility through partial and dynamic reconfiguration: http://wl.altera.com/devices/fpga/stratix-fpgas/stratix-v/overview/partial-reconfiguration/stxv-part-reconfig.html. Accessed 2016-04-11
5. Baranov SI (1994) Logic synthesis for control automata. Kluwer Academic Publishers, Boston, MA
6. Barkalov A, Wegrzyn M, Wiśniewski R (2006) Partial reconfiguration of compositional microprogram control units implemented on FPGAs. In: Proceedings of IFAC workshop on programmable devices and embedded systems (Brno), pp. 116–119
7. Batlle J, Martı J, Ridao P, Amat J (2002) A new FPGA/DSP-based parallel architecture for real-time image processing. Real-Time Imaging 8(5):345–356
8. Bukowiec A, Doligalski M (2013) Petri net dynamic partial reconfiguration in FPGA. In: Computer aided systems theory-EUROCAST. Springer, pp. 436–443
9. Chodowiec P, Gaj K (2003) Very compact FPGA implementation of the aes algorithm. In: Cryptographic hardware and embedded systems-CHES 2003. Springer, pp. 319–333
10. Crookes, D, Benkrid, K, Bouridane A, Alotaibi K, Benkrid A (2000) Design and implementation of a high level programming environment for FPGA-based image processing. In: Vision, image and signal processing, IEE proceedings-, vol 147. IET, pp. 377–384
11. Czerwinski R, Kania D (2012) Area and speed oriented synthesis of FSMs for PAL-based CPLDs. Microprocess Microsyst Embed Hardw Des 36(1):45–61
12. De Castro A, Zumel P, García O, Riesgo T, Uceda J (2003) Concurrent and simple digital controller of an ac/dc converter with power factor correction based on an FPGA. IEEE Trans Power Electron 18(1):334–343
13. Deepakumara J, Heys HM, Venkatesan R (2001) FPGA implementation of MD5 hash algorithm. In: 2001, Canadian conference on electrical and computer engineering, vol 2. IEEE, pp. 919–924
14. Doligalski M, Bukowiec A (2013) Partial reconfiguration in the field of logic controllers design. Int J Electron Telecommun 59(4):351–356
15. Eguro K, Venkatesan R (2012) FPGA for trusted cloud computing. In: 2012 22nd international conference on field programmable logic and applications (FPL). IEEE, pp. 63–70
16. Fons F, Fons M, Cantó E, López M (2013) Real-time embedded systems powered by FPGA dynamic partial self-reconfiguration: a case study oriented to biometric recognition applications. J Real-Time Image Process 8(3):229–251
17. Hauck S, DeHon A (2010) Reconfigurable computing: the theory and practice of FPGA-based computation. Morgan Kaufmann
18. Johnston C, Gribbon K, Bailey D (2004) Implementing image processing algorithms on FPGAs. In: Proceedings of the eleventh electronics new zealand conference, ENZCon'04, pp. 118–123
19. Kania D, Kulisz J (2007) Logic synthesis for PAL-based CPLD-s based on two-stage decomposition. J Syst Softw 80(7):1129–1141
20. Lattice Semiconductor homepage: http://www.latticesemi.com. Accessed 2016-03-04
21. Łabiak G, Wegrzyn M, Muñoz AR (2015) Statechart-based design controllers for FPGA partial reconfiguration. In: XXXVI symposium on photonics applications in astronomy, communications, industry, and high-energy physics experiments (Wilga 2015), International society for optics and photonics, pp. 96623Q–96623Q
22. Maxfield C (2004) The design warrior's guide to FPGAs. Academic Press Inc., Orlando, FL
23. Maxfield M (2014) ASIC, ASSP, SoC, FPGA? What's the difference?. EE Times
24. Monmasson E, Cirstea MN (2007) FPGA design methodology for industrial control systems—a review. IEEE Trans Ind Electron 54(4):1824–1842

25. Moreno-Munoz A, Pallarés-López V, la Rosa D, González JJ, Real-Calvo R, González-Redondo M, Moreno-García I (2013) Embedding synchronized measurement technology for smart grid development. IEEE Trans Ind Electron 9(1):52–61
26. Nichols RK, Lekkas PC (2002) Wireless security. McGraw-Hill New York
27. Okada S, Torii N, Itoh K, Takenaka M (2000) Implementation of elliptic curve cryptographic coprocessor over gf (2m) on an FPGA. In: Cryptographic Hardware and Embedded Systems—CHES 2000. Springer, pp. 25–40
28. Park J, Hwang J-T, Kim Y-C (2005) FPGA and ASIC implementation of ecc processor for security on medical embedded system. In: Third international conference on information technology and applications, 2005. ICITA 2005, vol 2. IEEE, pp. 547–551
29. Pingree PJ (2010) Advancing NASA's on-board processing capabilities with reconfigurable FPGA technologies. INTECH Open Access Publisher
30. Shreejith S, Fahmy SA, Lukasiewycz M (2013) Reconfigurable computing in next-generation automotive networks. IEEE Embed Syst Lett 5(1):12–15
31. Tanaka H, Ohnishi K, Nishi H, Kawai T, Morikawa Y, Ozawa S, Furukawa T (2009) Implementation of bilateral control system based on acceleration control using FPGA for multi-DOF haptic endoscopic surgery robot. IEEE Trans Ind Electron 56(3):618–627
32. Vivado design suite tutorial: Partial reconfiguration: http://www.xilinx.com/support/documentation/sw_manuals/xilinx2016_1/ug947-vivado-partial-reconfiguration-tutorial.pdf. Accessed 2016-04-11
33. Vivado design suite user guide: http://www.xilinx.com/support/documentation/sw_manuals/xilinx2013_1/ug901-vivado-synthesis.pdf. Accessed 2016-04-11
34. Wegrzyn M, Adamski M, Karatkevich A, Rosado-Muñoz A (2014) FPGA-based embedded logic controllers. In: Proceedings of the 7th IEEE international conference on human system interactions. Lisbon, Portugal, pp. 249–254
35. Wiśniewski R (2009) Synthesis of compositional microprogram control units for programmable devices, vol 14. Lecture Notes in Control and Computer ScienceUniversity of Zielona Góra Press, Zielona Góra
36. Wiśniewski R, Barkalov A, Titarenko L (2008) Partial reconfiguration of compositional microprogram control units implemented on an FPGA. In: Proceedings of IEEE East-West design & test symposium-EWDTS, vol 8, pp. 80–83
37. Wiśniewski R, Grobelna I, Stefanowicz Ł (2016) Partial reconfiguration of concurrent logic controllers implemented in FPGA devices. In: 12th Internatinal conference of computational methods in sciences and engineering—ICCMSE'16, pp. TBD., Athens, Greece. (accepted for publication)
38. Wiśniewski, R, Wiśniewska M, Adamski M (2016) Effective partial reconfiguration of logic controllers implemented in FPGA devices. In: Design of reconfigurable logic controllers. Springer, pp. 45–55
39. Xilinx homepage: http://www.xilinx.com. Accessed 2016-03-04
40. Xilinx implementation overview for FPGAs: http://www.xilinx.com/support/documentation/sw_manuals/xilinx11/ise_c_implement_fpga_design.htm. Accessed 2016-04-11
41. Xilinx partial reconfiguration user guide: http://www.xilinx.com/support/documentation/sw_manuals/xilinx14_1/ug702.pdf. Accessed 2016-04-11
42. Xilinx Spartan 3E Data Sheet: http://www.xilinx.com/support/documentation/data_sheets/ds312.pdf. Accessed 2016-04-11
43. Xilinx Virtex-II Pro Data Sheet: http://www.xilinx.com/support/documentation/data_sheets/ds083.pdf. Accessed 2016-04-11
44. Xilinx Virtex-5 User Guide: http://www.xilinx.com/support/documentation/user_guides/ug190.pdf. Accessed 2016-04-11

Chapter 10
Conclusions

Prototyping of control systems is almost always related to concurrency. For example, execution of tasks in traditional computers, phones, or tablets are usually split into multi-threads, computed by several microprocessors. However, proper designing of controllers oriented on concurrency involves application of advanced algorithms and techniques in order to prepare the system.

This book proposes the complete prototyping flow of concurrent control systems implemented in the FPGA. The system is initially specified by an interpreted Petri net, which naturally reflects concurrency and sequentiality relations in the design. Besides the traditional analysis methods (liveness, safeness), additional techniques are proposed. Chapter 5 shows the known and introduces new analysis algorithms regarding concurrency and sequentiality.

Further decomposition of the system splits the controller into sequential automata. Algorithms proposed in Chap. 6 can be applied depending on the required needs. For example, linear algebra and hypergraphs theory assure the minimal possible number of decomposed components, however exponential complexity of such methods may lead to the situations, where the solution cannot be found. On the other hand, the application of comparability graphs reduces the computational complexity to polynomial, but not all Petri nets can be decomposed with the use of this technique.

Once the system is decomposed, the achieved components ought to be properly synchronized. Since the whole controller is implemented in a single FPGA device, all the modules operate in the same time domain (that is, use the same clock signal). Therefore, the synchronization algorithm presented in Sect. 7.2.2 is universal and can be applied to any integrated concurrent control system prototyped according to the guidelines proposed in this book.

Modelling of the decomposed components can be performed in various ways. Section 8 proposes the application of traditional finite state machines, but any form of sequential automata can be used. The modelled components are described with the use of hardware languages, such as Verilog or VHDL.

© Springer International Publishing Switzerland 2017

R. Wiśniewski, *Prototyping of Concurrent Control Systems Implemented in FPGA Devices*, Advances in Industrial Control,
DOI 10.1007/978-3-319-45811-3_10

Finally, the system is logically synthesized and implemented. These steps are strictly executed according to the requirements of the vendor of the supplied device. In the descriptions shown in Chap. 9, *Xilinx* FPGAs were selected as representative.

The main benefits of the proposed prototyping flow are emphasized in Sect. 9.4, where innovate techniques of partial reconfiguration of concurrent controllers are shown. The complete bit-stream of the system is downloaded to the FPGA only once. Further modifications to the controller are done with the use of partial bit-streams which means, that only a portion of data is sent to the FPGA. Two reconfiguration techniques are proposed in the book. The first one, called *static* partial reconfiguration allows easy and comfortable reconfiguration of any modules of the prototyped system. However, the controller cannot perform any action during the configuration process. *Dynamic* partial reconfiguration requires additional modifications to the prototyped system, as it is shown in Sect. 9.4.4, but the benefits are much more fruitful. The part of the system is reconfigured, while the remaining modules of the controller are still working, executing the assigned tasks.

In the author's opinion, the main contributions and results presented in this book are the following:

- Formulation of the novel prototyping flow of the concurrent control system oriented on further *dynamic* partial reconfiguration.
- Formulation of the novel prototyping flow of the concurrent control system oriented on further *static* partial reconfiguration.
- Formulation of the novel decomposition algorithm of concurrent control system described by an interpreted Petri net based on comparability graphs. The method includes supplementation of the decomposed net by nonoperational places (NOPs). The main advantage of the presented method is its polynomial computational complexity, however not all the systems can be decomposed with the application of comparability graphs.
- Formulation of novel methods regarding concurrency and sequentiality analysis in the system described by an interpreted Petri net.

Looking into more details, the contributions regarding algorithms (with their computational complexities estimation), theorems, lemmas and proofs can be summarized as follows:

- Formulation of the novel algorithm for simultaneous recognition and coloring of comparability graphs, supplemented by adequate theorems, lemma and proofs regarding its computational complexity (cf. Sect. 3.4: Algorithm 3.2, Theorems 3.2, 3.3 and Lemma 3.1).
- Formulation of formal algorithms, novel theorems and proofs regarding exact transversals and c-exact hypergraphs (cf. Chap. 4: Algorithms 4.1 and 4.2, Theorems 4.1, 4.2 and 4.3).
- Formulation of novel theorems regarding the analysis of concurrency and sequentiality in a concurrent control system described by an interpreted Petri net (cf. Chap. 5: Theorems 5.4, 5.6, 5.7, 5.8, 5.9, 5.10, 5.11 and 5.12).

- Formulation of the novel synchronization algorithm for a decomposed concurrent control system intended for further implementation in a single device (decomposed components are oscillated by the same clock signal, cf. Chap. 7: Algorithm 7.1).

Certainly, most of prototyping techniques presented in this book are not limited to controllers implemented in the FPGA. The proposed methods and algorithms combine various scientific areas: distributed systems, electronics (FPGAs and programmable devices), computer science (algorithms for soft computing), mathematics (perfect graphs, hypergraphs, Petri nets).

For example, the analysis and decomposition methods shown in Chaps. 5 and 6 can be easily adopted to the prototyping flow of distributed controllers. Such systems require decomposition in order to split the system into the components located over various devices (such as computers, microprocessors, programmable devices, etc.). Clearly, the decomposition methods shown in the book can be used in such cases. However, synchronization between the decomposed components usually requires different techniques. In case of a distributed control system, particular modules are oscillated by different clocks, thus they work in different time domains.

The proposed algorithms for partial reconfiguration (Chap. 9) can be used for any system that is implemented in FPGA. Of course, the prototyping technique may be different. However, the main reconfiguration idea remains the same. Especially *dynamic* partial reconfiguration requires similar changes to the system (splitting into static and reconfigurable parts).

Finally, algorithms regarding comparability graphs and c-exact hypergraphs can be applied in various disciplines beside computer science and electronics (mainly mathematics, biology, chemistry). Unique properties of such structures permit solving some tasks polynomially that are exponential in general case. It seems that the development of the presented algorithms may lead to further improvements and enhancements.

Index

© Springer International Publishing Switzerland 2017
R. Wiśniewski, *Prototyping of Concurrent Control Systems Implemented in FPGA Devices*, Advances in Industrial Control,
DOI 10.1007/978-3-319-45811-3

MIX
Papier aus verantwortungsvollen Quellen
Paper from responsible sources
FSC® C105338

If you have any concerns about our products,
you can contact us on
ProductSafety@springernature.com

In case Publisher is established outside the EU,
the EU authorized representative is:
Springer Nature Customer Service Center GmbH
Europaplatz 3, 69115 Heidelberg, Germany

Printed by Libri Plureos GmbH
in Hamburg, Germany